ALFRED WEGENER

THE FATHER OF CONTINENTAL DRIFT

SCIENTIFIC REVOLUTIONARIES:
A Biographical Series

MARTINUS WILLEM BEIJERINCK
His Life and Work
By G. van Iterson, Jr., L.E. den Dooren de Jong, and A.J. Kluyver

Forthcoming books in this series

OTTO HAHN
A Scientist of Our Times
By Walther Gerlach and Dietrich Hahn

ROBERT KOCH
His Life and His Work
By Thomas D. Brock

ALFRED WEGENER

THE FATHER OF CONTINENTAL DRIFT

By Martin Schwarzbach

With an Introduction to the English edition
by Anthony Hallam

And an Assessment of the Earth Science Revolution
by I. Bernard Cohen

SCIENCE TECH, INC.
Madison, Wisconsin 53705

Originally published as *Alfred Wegener und die Drift der Kontinente*

Science Tech, Inc., 701 Ridge St.,
Madison, Wisconsin 53705 USA

Translation: Carla Love
Production supervision: Science Tech, Inc.
Editorial supervision: Ruth B. Siegel
Interior design: Thomas D. Brock
Cover design: Katherine M. Brock

Library of Congress Cataloging-in-Publication Data

Schwarzbach, Martin, 1907–
Alfred Wegener, the father of continental drift.

(Scientific revolutionaries)
Translation of: Alfred Wegener und die Drift der Kontinente.
Bibliography: p.
Includes index.
1. Wegener, Alfred, 1880–1930. 2. Earth scientists—Germany—Biography. 3. Continental drift. 4. Earth sciences—History. I. Title. II. Series.
QE22.W26S3913 1986 551.1'36 86-13760
ISBN 0-910239-03-7

Printed in the United States of America

10 9 8 7 6 5 4 3 2 1

Contents

Alfred Wegener.

Introduction to the English Edition

Alfred Wegener is a romantic figure in the history of science. His story is one of the outsider who early in this century proposed a theory which proved to be too radical for most of his more knowledgeable contemporaries. Nevertheless, he was vindicated in essential matters many years after his death by the so-called earth sciences revolution, so that today the idea of continental mobilism is almost universally accepted. Such a dramatic story provokes many questions: Why were Wegener's arguments not accepted earlier? Did his contemporaries have what seemed to them to be telling counter-arguments, or were they the victims of conservative prejudice or blinkered vision? Or were critical data simply lacking? Why did such a scientist emerge in Germany rather than in some other country where geological research was more active? Was this a matter of chance or was the intellectual environment in Germany more conducive to mobilistic thoughts than elsewhere? How good a scientist was Wegener and what was he like personally?

Enough has been written to allow a preliminary answer to some of these questions, but until this translation of Professor Schwarzbach's book appeared, there was no substantial account in the English language of Wegener's life and work. The fullest biographical account remains that of his widow, available only in German, which is an invaluable source for all kinds of documentation. Schwarzbach distils the essence of this work and adds a good deal more, on the basis of his experience as a distinguished geologist in his own right, who has made important contributions to paleoclimatology, a subject pioneered by Wegener and Wegener's father-in-law Wladimir Köppen. Writing his book in time to appear during the centenary of Wegener's birth in 1980, Schwarzbach was also in a good position to reflect on the earth sciences revolution a few years earlier that had led to the emergence of plate tectonics as the dominant paradigm, and to assess the role of Wegener's continental drift hypothesis in this development.

The centenary year was used by the German geological community as the occasion to acknowledge formally, and for the first time to a world audience, the great contribution of a native son who had recently become celebrated as a world figure. This was achieved by holding an international conference in West Berlin, the proceedings of which were published in *Geologische Rundschau*, volume 70. During this conference, at which distinguished overseas guests reviewed the current state of the art in their particular research fields, indicating how they had been enriched by plate tectonic theory, speaker after speaker testified to the logical thought, sharp insight, and prescience that had characterized so much of Wegener's continental drift work.

This would all have been something of a surprise to the many Germans who mourned Wegener's death on the Greenland ice cap half a century earlier. His laudatory obituaries concentrated on the domain for which he was best known during his lifetime—namely his courage, skill, and scientific flair—as an explorer on four expeditions to his beloved Greenland. As a scientist he was remembered principally as a pioneering meteorologist, for which his Greenland work was essential, leading among other things to the earliest recognition of the jet stream in the upper atmosphere. It is entirely appropriate therefore that Schwarzbach devotes full attention to these subjects, which occupied most of Wegener's professional life. He was so busy, indeed, in the latter part of the 1920s preparing for his final Greenland expedition, that he had no time to reply adequately to the increasing amount of opposition to his theory of continental drift. This opposition followed the translation of his seminal book *The Origin of Continents and Oceans* into many languages, which allowed Wegener's ideas to become widely known and evaluated.

Schwarzbach also reminds us of the incredible energy, courage, and endurance required of polar expeditioners earlier this century, without today's technological support, and his account of Wegener's last trips across the Greenland ice cap makes harrowing reading. So much for those who contemptuously dismissed Wegener as a mere armchair theorist. Any polar expeditioner requires other essential personal qualities to enable him to get on with his companions in all kinds of intimate and awkward circumstances, as well as a strong organizational capacity. It is evident that Wegener was endowed with the requisite qualities in full measure. His straightforward personality, equable temperament, and dry sense of humor, together with his single-minded devotion to the ideals of scientific exploration, would have made him someone to cherish as companion and friend.

There are many revealing morsels to savor in Schwarzbach's book. Thus today's geology students may be reassured that Wegener had no great love of or ability in mathematics. Also, it is quite clear from early correspondence that he did not rely on matching of modern coastlines to achieve the celebrated Atlantic "jigsaw fit," as some of his critics alleged. On the other hand, during his visit to Iceland he failed to appreciate the significance of the abundant evidence for tensional tectonics associated with volcanicity, which would have lent strong support to his continental drift theory. Schwarzbach also reminds us of the contributions of other geological theorists in the German-speaking world early this century. While Ampferer's ideas on mantle convection currents anticipate later thought, the dominant influence of Stille's stabilist geotectonics inhibited many geologists from taking Wegener's ideas more seriously.

Thus, Schwarzbach's observations stimulate our desire for more information, especially about the contributions to geotectonics of early twentieth century geologists in continental Europe. There is, indeed, abundant scope for an in-depth scholarly analysis of Wegener in the context of contemporary thought, along the lines pursued by Mott Greene in his excellent book *Geology in the Nineteenth Century* which revealed the emergence of quite different traditions of geological research in continental Europe, Great Britain, and the United States. Meanwhile we may be grateful to Professor Schwarzbach for providing us with such a readable and illuminating memoir.

A. Hallam
Lapworth Professor of Geology
University of Birmingham
Birmingham

PREFACE

For everyone interested in the earth sciences, Alfred Wegener's life is a fascinating story. Wegener was a remarkably versatile and creative scientist. He wrote his dissertation on a topic in astronomy. In his professional life he was primarily a meteorologist, and his textbook, *Thermodynamik der Atmosphäre*, which was written when he was thirty, went through several editions. In Graz, Austria, he was professor of meteorology and geophysics. However, from boyhood on his goal had always been to explore the Arctic, and he did, in fact, participate in four expeditions to Greenland, each of which demanded extraordinary amounts of physical endurance, "iron will, and energy." Wegener's explorations of Greenland also became well known to the general public though his entertaining accounts of his travels. Another feat of physical endurance—although perhaps more trivial—was the world record in ballooning set by Wegener and his brother Kurt, when they were in their twenties.

His true and lasting importance, however, is found in an entirely different field. In 1912 he published, for the first time, his revolutionary hypothesis that the continents are not fixed but rather have been slowly wandering during the course of earth history. This idea, which became known as "continental drift," was not completely new, but Wegener was the first to develop it in depth.

This hypothesis made it possible to solve many different geological and geophysical problems. Seldom has a new set of ideas about the origin of the earth's crust, mountain ranges, and oceans had such a revitalizing effect as did Wegener's theory of continental drift. After Wegener's hypothesis, climatic changes in geologic time now also appeared in a new light: In this field Wegener's collaborative work with his father-in-law, the climatologist Wladimir Köppen, would be particularly important.

A half-century after Wegener's death, and seventy-five years after his formulation of the theory of continental drift*, it seems appropriate to undertake a new summation of his life and discussion of his importance as a scientist. Of course, we already have the detailed German biography written in 1960 by his wife, Else. She was the person who knew best this brilliant but modest man who lived only for his research, and she had a wealth of letters and diaries at her disposal. We also have the many obituary notices and memorial essays—of varying length and quality—written by his colleagues. In all of these, however, the spotlight is on Wegener, the explorer and meteorologist. Since 1912, Wegener the scientist has been of interest mainly to geologists and to geophysicists working in areas closely related to geology. Thus we are justified in taking a new look at Wegener's life and his real life's work from this perspective, especially since in the intervening years a large volume of new data, which were completely unsuspected at Wegener's time, has become available. A chapter will also be devoted to his work in Iceland.

In many respects, the verification of Wegener's hypothesis of continental drift was as problematic as its formulation was

*The term "continental drift" did not become popular in the English language until the 1920's (Hallam, 1973), although Wegener first published his revolutionary hypothesis in 1912. Wegener himself used the German word "Verschiebung" (English, "displacement") and used the word "drift" in later publications only after the term had received wide usage in English. In the present book, for simplicity, only the term "drift" is used.

Else and Alfred Wegener in Marburg in the winter of 1913.

revolutionary. A number of geological and geophysical arguments in its favor turned out to be unsound, and during his lifetime Wegener was unable to win acceptance for his theory, either in Europe or America—not least because as an "outsider" he was not taken seriously by many geologists. Only after World War II did completely new data—especially from modern deep-sea exploration and from research into the earth's ancient magnetic field—lead to a more coherent synthesis, "plate tectonics", which, for the last twenty years, has dominated discussion of the continents and oceans, earthquakes and volcanoes, and the formation of the great mountain chains. Although many details of this theory, too, are disputed, it provides a much more consistent and complete picture than Wegener was able to sketch. As far as the end result is concerned, the correctness of Wegener's theory of continental drift can be affirmed today: parts of the earth crust are moving. On that point Wegener's hypothesis and the hypothesis of plate tectonics coincide. But otherwise they are two completely different theories.

At first, Wegener was ridiculed as a "teller of tall tales"; in the end he was compared to the great Copernicus. That comparison is not completely justified; but as a "revolutionary" in the realm of the earth sciences, Wegener certainly takes precedence over his much more-famous German compatriot Alexander von Humboldt—although that judgment, which is perhaps here expressed for the first time, is not intended in the least to diminish Humboldt's legendary renown.

The present study is not a textbook about continental drift and plate tectonics; it goes into only the most important fundamentals of these concepts. A number of them are explained in some detail, often in the notes; the scientist to whom they are already familiar can easily skip them, but the general reader will find them helpful. Thus this book is intended for everyone who is interested in the history of the earth and in the last one hundred years of research into that history. Particular attention will be given to the importance to ice age research of Wegener's (and Köppen's) bold ideas—from the Gondwana glaciations of the Permocarboniferous to the "radiation curves" of Milankovitch and the "next Ice Age."

To all of those who provided help and information I would like to express my thanks, especially to Frau Else Wegener, Alfred Wegener's wife and lifelong companion, who most graciously answered all my questions. Many letters and other documents referring to Wegener were lost in the chaos of World War II and the post-war years. Frau Wegener was able to donate some materials to the archives of the Deutsches Museum in Munich, to the historical museum in Neuruppin and to the museum at the Wegener Memorial in Zechlinerhütte, thus making them more generally accessible.

Chronology of Alfred Wegener's life

1880, November 1	Born in Berlin.
1899	Completed secondary education at *Cöllnischen Gymnasium*, Berlin.
1899–1904	Attended the Universities of Heidelberg, Innsbruck, and Berlin, studying the natural sciences, especially astronomy.
1904, November 24	Received Ph.D., University of Berlin.
1905–1906	Assistant at Aeronautic Observatory in Lindenberg.
1906–1908	Participated in Danish Greenland expedition led by Mylius-Erichsen.
1909	Qualified to teach meteorology and astronomy at the University of Marburg (*Habilitation*).
1909–1919	*Privatdozent* and, after 1917, professor at the University of Marburg.
1911	*The Thermodynamics of the Atmosphere* (second edition, 1924; third edition, 1928).
1912, January 6	First presentation of the drift hypothesis at the *Geologischen Vereinigung*, Frankfurt.

1912–1913	Participated in Danish Greenland expedition led by J. P. Koch; traversed Iceland and Northern Greenland.
1913	Married Else Köppen, daughter of the well-known meteorologist Wladimir Köppen.
1914–1918	Served in World War I.
1915	First edition of *The Origin of Continents and Oceans* (second edition, 1920; third edition, 1922; fourth edition, 1929).
1919–1924	Department head at the German Marine Observatory in Hamburg; professor at the University of Hamburg.
1922	Measured upper air currents over Atlantic Ocean during voyage to Cuba and Mexico.
1924	*The Climates of the Geologic Past* (with Wladimir Köppen).
1924–1930	Professor of geophysics and meteorology at the University of Graz, Austria.
1926	Continental Drift Symposium held in New York (which Wegener did not attend.)
1929	Led preparatory expedition to Greenland.
1930	Led German Greenland expedition.
1930, November	Died while crossing Greenland's ice cap.

1

ALEXANDER VON HUMBOLDT AND THE ANCESTORS OF ALFRED WEGENER

How quickly last winter flew by—and how long this one seems! God, what happy hours we all chatted away around your stove, next to your old worn chair! Not a day went by that we didn't see each other once or twice. Now we're all scattered—where are our old friends?

—Alexander von Humboldt[1]

Alexander von Humboldt and the Ancestors of Alfred Wegener

A biography of Alfred Wegener can begin with Alexander von Humboldt—the German naturalist, explorer, and statesman (1769–1859), who played an important role in the development of European physical geography in the nineteenth century. Humboldt was best known for his discovery of the ocean current off the west coast of South America which was later named after him. Soon after Alexander von Humboldt began his studies at the University of Frankfurt-an-der-Oder in the winter semester of 1787–88, at the age of eighteen, he became acquainted with the theology student Wilhelm Gabriel Wegener. "What meant most to me was my friendship with Alexander von Humboldt, who became so famous later on," wrote Wilhelm Wegener in his unpublished autobiography.[2] "Alexander sought out my friendship with such genuine warmth and courtesy as no one had ever before shown me. He spent many hours with me, almost to the exclusion of everyone else. He is an estimable man and, to me, a valued friend."

Their bond of friendship is evident in the letters—highly emotional by today's standards—that they exchanged between 1788 and 1790, during the period of European Romanticism. Their paths had parted in the summer of 1788. Wilhelm Wegener continued his study of theology in Frankfurt-an-der-Oder, while Humboldt went to Berlin, Göttingen, and finally to Freiburg in Saxony to study with Abraham Gottlob Werner, the

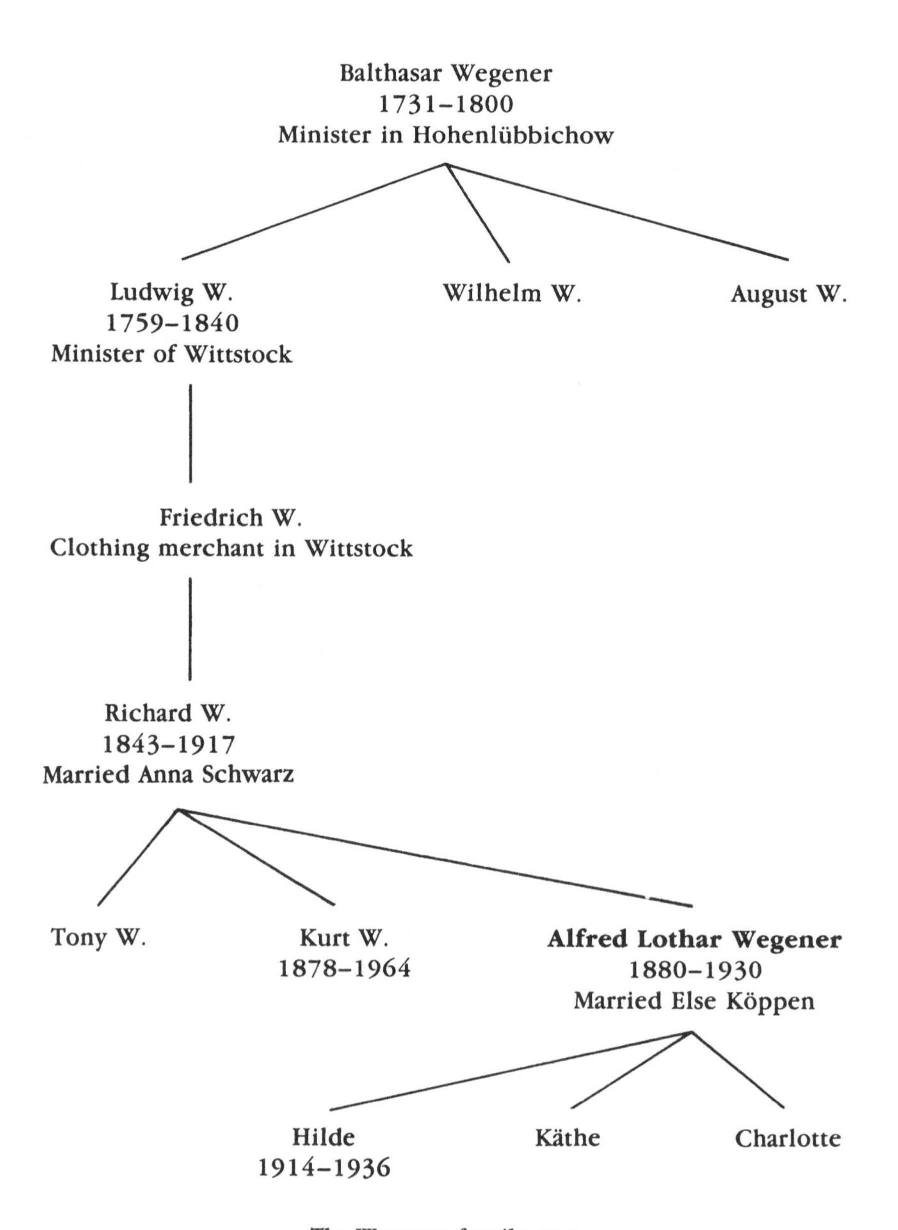

The Wegener family tree.

"father" of German geology,[3] as his interests turned more and more away from "wretched public administration" and toward the natural sciences. "Fortunately, I saved most of his letters to me, which attest to the clarity of his thought and his friendship for me," Wilhelm Wegener wrote. These letters were last published in 1973.[4] Humboldt, on the other hand, admitted late in his life to "cheerfully burning" most of the many letters written to him, beginning with those written to him by Goethe and Schiller.

Alexander von Humboldt's and Wilhelm Wegener's correspondence, so lively at first, soon came to an end. The last surviving letter to Wegener was written from Hamburg on September 23, 1790, after Humboldt's return from a trip to England with Georg Forster, the influential German scientist and writer.[5] Humboldt had been inundated by so many new ideas, especially in the natural sciences, that the theologian Wilhelm Wegener was no longer the right correspondent for him. In 1790 he wrote to him, "I am sending you my book about the basalts. You will be doing me a favor if you read the first half; it is entirely philological. The rest will bore you."[6]

Wilhelm Wegener, who was born in Hohenlübbichow in Brandenburg in 1767, became an army chaplain in the Regiment Gendarmes in Berlin upon completing his study of theology in 1789, and in 1795 he became senior pastor in Züllichau, where he died in 1837. He shares not only his rather common surname with Alfred Wegener. In fact, he belongs to his immediate ancestral line, since Wilhelm's brother Ludwig was Alfred Wegener's great-grandfather. The short, abbreviated family tree shown here illustrates these familial relationships for the last 250 years.

We will have occasion to come back to Alexander von Humboldt in a completely different context, namely as one of the scientists who took note of the remarkable shape of the Atlantic Ocean. From that same observation sprang Alfred Wegener's theory of continental drift.

2

ALFRED WEGENER'S LIFE AND CAREER

Vita nostra brevis est,
brevi finietur,
venit mors velociter,
rapit nos atrociter,
nemini parcetur.

Our life is brief,
It passes in a moment.
Death comes quickly,
It seizes us cruelly,
Sparing no one.

—"GAUDEAMUS IGITUR"[1]

ALFRED WEGENER'S LIFE AND CAREER

Berlin and the university years (1880–1904)

Pereat tristitia,
pereant osores,
pereat diabolus,
quivis antiburschius,
atque irrisores!

Let sadness vanish,
Let those who hate, perish.
Also the devil,
All those opposed to students,
And scoffers, too!

—"GAUDEAMUS IGITUR"[1]

Unfortunately, we do not know very much about Wegener's childhood and youth. However, we do have the *vita* which he was required to write in 1904 as a supplement to his doctoral dissertation.

> I, Alfred Lothar Wegener, a Protestant, was born in Berlin on November 1, 1880, the son of Dr. Richard Wegener, a pastor and the director of the Schindler Orphanage. I received my secondary education at the *Cöllnischen Gymnasium* in Berlin, from which I graduated in September 1899.[2] I then went on to the Friedrich Wilhelm University in Berlin, where I began my study of mathematics and the natural sciences, astronomy in particular. Apart from the summer semesters of 1900 and 1901, which I spent at the

> Ruprecht Karl University in Heidelberg and the University of Innsbruck, respectively, I completed my studies in Berlin.
>
> From September 1901 to September 1902 I fulfilled my year of required military service as a volunteer in the Queen Elizabeth Guard-Grenadier Third Regiment in Westend. From September 1902 to September 1903 I worked as an astronomer at the observatory of the Urania Society [in Berlin]. I passed my doctoral examination on November 24, 1904. In the ten semesters from September 1899 to September 1904, I attended courses taught by the following professors: Bauschinger, von Betzold, Blaas, Cathrein, Dilthey, Eggert, Fischer, Förster, Frobenius, Fuchs, Heinricher, Helmert, Knoblauch, Königsberger, Markuse, Paulsen, Planck, Quincke, Scheiner, Schwarz, Stumpf, Valentiner, Warburg, and Wolf. From my seventh semester on I participated in the seminars given by Professors Bauschinger and Förster, to whom I am especially grateful for their good advice and valuable suggestions.

Alfred's father, Richard Wegener, had studied Protestant theology and classical languages and received his Ph.D. from the University of Berlin. The Schindler Orphanage, where he served as director from 1875 to 1904, was a private institution for the sons of civil servants, teachers, and clergymen, who concurrently attended a highly regarded secondary school, the *Gymnasium zum Grauen Kloster*. (The school's most prominent graduate was Otto von Bismarck, the famous Prussian leader, who graduated in 1832.) Dr. Wegener also taught at the *Gymnasium*, but he sent his sons Alfred and Kurt to a different school, the *Cöllnischen Gymnasium*, which gives some idea of his strict adherence to principle.

Alfred was the youngest of five children, two of whom died in childhood. His sister Tony and his brother Kurt,[3] two years his elder, survived him. The metropolis of Berlin was not the only locale to put its stamp on their childhood. Even more important was the family's vacation home in Zechlinerhütte, a small village in the charming landscape of hills and lakes in

In 1980 this memorial tablet in honor of Alfred Wegener was placed on the wall of the high school he attended in Berlin. The inscription reads, "The polar investigator Alfred Wegener was a student in this building, formerly the Cöllnischen Gymnasium; *he graduated from the University of Berlin in 1905; he laid the foundation for modern geology with his theory of continental drift."*

The former residence of the manager of the Zechlinerhütte glass foundry and the vacation and retirement home of Wegener's parents, now a Wegener Memorial. Photographed in 1978.

the northern part of Germany near Rheinsberg, some 90 kilometers north of Berlin. Since 1737 a crystal glass foundry was in operation there; it was finally shut down in 1889. In 1886 Wegener's father bought the foundry manager's modest house, and this became the family's permanent vacation residence. There were also old family ties to Zechlinerhütte, for Richard Wegener's wife, Anna (née Schwarz), had been born there. Richard and Anna had met in nearby Wittstock, where Richard's father was born and where Anna, who was orphaned in childhood, was raised.

This landscape, which still bore the imprint of the Ice Age in its deep forests and quiet lakes, was an ideal complement to the "inseparable" brothers' natural bent for the sciences and their love of nature. It is not surprising that as they grew older, they were always glad to come back to Zechlinerhütte, which eventually became their parents' retirement home. The last two to bear the family name—the restless, widely travelled Kurt and his sister Tony—neither of whom ever married, are buried there next to their parents[4]. Today the family home in Zechlinerhütte is a Wegener Memorial.

After completing their secondary education, both Kurt and Alfred turned to the study of the natural sciences and thus departed from the old family tradition. In some respects their lives ran a similar course: for both of them meteorology was an important focus professionally, and for both of them that interest was linked with their enthusiasm for exotic travels. For several years they worked together in Lindenberg and Hamburg; and after Alfred's death in Greenland, Kurt led the final phase of the expedition and then assumed for a time his late brother's professorship in Graz. For the most part, however, their professional paths diverged. In physical appearance the two were completely different. Else Wegener wrote of her husband and her brother-in-law, "Everything about Kurt was slim and slender, and he was quite a bit taller than Alfred. Kurt's slender hands would fumble nervously in his pockets as he looked for his ticket. How composed Alfred seemed by contrast! . . . With his medium build Alfred appeared muscular and strong."[5]

Alfred spent his first summer semester in Heidelberg. If one thinks of Wegener the Greenland explorer, totally committed to his scientific work and to the success of his expedition despite nearly insurmountable difficulties, one would not suspect that the young astronomy student in Heidelberg was an enthusiastic member of a traditional student club, or that during that summer he never entered the famous Königsstuhl Observatory. (Wegener's Heidelberg transcript indicates that he did register for the course "Elements of Meteorology," taught by Max Wolf,[6] the well-known astronomer who was director of the observatory, and for Wilhelm Valentiner's "General Astronomy.") The transition from the home of his loving, but strict parents to the free and easy life of a student in a romantic university town may well have led the young Wegener, as his wife wrote later, "to sow all his wild oats at once." An amusing document from that summer is the following police citation:

> The accused, Alfred Wegener, a student, residing at Heinrich Ingrimmstrasse 5, is charged with gross misconduct and disturbing the peace, in that on the third of this month at 3 a.m. he paraded down the main street to the market square wrapped in a white sheet and shouting loudly in an unseemly manner.
>
> —Policeman Eiermann
>
> In accordance with Paragraph 360 of the Civil Code he is assessed a fine of 5 marks plus costs. In case of non-payment, the penalty will be two days incarceration.
>
> —Heidelberg, July 3, 1900.

During his university years in Berlin, Wegener was a member, at least for a time, of the "Academic Astronomy Club," a scientific society for students. In his memoirs the mathematician Walter Lietzmann[7] describes his fellow club member Alfred Wegener, "who even then dreamed of the Arctic and liked to sing the song about the Eskimo man and woman." That was probably around 1900.

That Wegener kept his sense of humor later on in life, even if his natural restraint did not always let it show, is evident from his wife's descriptions of him, as well as from the charming memorial essay written by Hans Benndorf, his friend and colleague in Graz, and from many passages in his letters and diaries.

Wegener also attended the University of Innsbruck for one summer semester together with his brother Kurt. As is true for many of the non-Austrian students who come to this Tyrolean university town, the real attraction for the two brothers was the Alps and the wonderful opportunities they offered for many different, and sometimes difficult, hikes and climbs—apt preparation for Wegener's future plans, although these were, of course, still unformed.

Wegener completed the major part of his studies at the University of Berlin. The *vita* in his dissertation includes the customary list of professors whose lectures he attended, but the list tells us nothing of the quality or accomplishments of the courses. Max Planck,[8] who would later win a Nobel Prize, and Emil Fischer, who had won the Prize already, are among those listed; it would be interesting to know what sort of impression they made on Wegener the student. Besides meteorology, which he studied with Wilhelm von Bezold,[9] astronomy was foremost in his interests, as his dissertation topic makes clear: "The Alfonsine Tables for the Use of a Modern Calculator." As dissertation advisors, Professors Bauschinger[10] and Foerster[11] are named. Bauschinger's primary interest was in determining the orbits of celestial bodies, and for a long time he was the head of the Institute for Astronomic Calculation in Berlin. Wegener's topic fit in perfectly with this line of research. On November 24, 1904 Wegener passed his doctoral examination *magna cum laude.*

The Alfonsine Tables are an old set of tables which were used to calculate the positions of the sun, the moon, and the five planets known at that time. They were developed in the thirteenth century—the age of the Ptolemeic system—at the instigation of Alfonse X of Castille,[12] and were indispensable for sea travel, even though they were not always accurate and, moreover, were difficult to use. On that point Alphonso "the Wise" is supposed to have remarked, "If God had consulted me when he created the world, I would have recommended greater simplicity."[13]

Wegener recalculated the tables for the "use of a modern calculator," just as Johannes Kepler had once recalculated Tycho

Brahe's earlier observations in his "Rudolphine Tables" on the basis of his new findings about planetary motion. This publication of Wegener's, his first, is thus purely mathematical, and remained his only one pertaining exclusively to astronomy. Even his later investigations of astronomical objects, such as the Hessian meteor of 1916, or of the origin of the lunar craters, focus on related fields such as geology. "In astronomy everything has essentially been done. Only an unusual talent for mathematics together with specialized installations at observatories can lead to new discoveries; and besides, astronomy offers no opportunity for physical activity." In this view of Wegener's, which Benndorf quotes in his memorial essay of 1931, we have the reasons that he rejected astronomy as a career.

Lindenberg (1905–1906)

One must be especially grateful to the Director of the Aeronautic Observatory, Richard Assmann, for having recognized the special talents of the brothers and for giving them positions at the observatory. They came to the institute during that exciting time when the main task was to improve the technology of kite and balloon ascensions, a task for which creative people were always needed.[14]

Wilhelm Foerster, already mentioned earlier, was an energetic, administratively able astronomer, who had founded the Urania Public Observatory in Berlin in 1889. Wegener worked there as an assistant during his last few semesters at the university. But after he completed his studies he gave up this astronomical activity and turned his complete attention to practical meteorology. Along with his brother Kurt he became a "technical aide" at the Aeronautic Observatory in Lindenberg,[15] in the district of Beeskow (a suburb of Berlin). The brothers were probably attracted by the opportunity to take part in modern atmospheric research, using kites and balloons, and to ascend in balloons themselves. At the Wegener Memorial in Zechlinerhütte, one can see Wegener's membership card for the Fédération Aéronautique Internationale, validated in 1911 and signed by Franz Richarz, the physics professor at Marburg.[16]

How much the physical side of their work appealed to them can be seen from the balloon journey they made from April 5 to April 7, 1906. They went from Bitterfeld in central Germany, to Jutland in northern Denmark, and then south to the Spessart region in central Germany, east of Frankfurt-am-Main. This was a feat which attracted a good deal of public attention at the time. As far as food and clothing were concerned, the brothers were poorly equipped for a trip of that distance, and it was only thanks to their tenacious endurance and meteorological skill that they were able to stay aloft for fifty-two hours—they broke the previous world record by seventeen hours!

In 1906, Wegener's work in Lindenberg was interrupted for two years by the unexpected opportunity to take part in the Danish "Danmark" expedition to Greenland, under Ludvig Mylius-Erichsen. Wegener's exploration of Greenland is dealt with in a later chapter, so that we need not go into detail here. Another event of 1906 also turned out to be of deciding importance in Wegener's life: his first contact with Wladimir Köppen, a highly regarded meteorologist and the head of the meteorological kite station at Grossborstel near Hamburg, whom Wegener asked for advice about the research he planned to do in Greenland. In 1908, after his return from the Arctic, he visited Köppen's home and became acquainted with his sixteen-year-old daughter, Else, whom he married five years later.

Up, up, and away—the Wegener family aloft, April 17, 1912. From left to right: Alfred Wegener, Kurt Wegener, Else Köppen, Tony Wegener.

Marburg (1908–1918)

> *The war found me in Marburg. One day a man visited me whose fine features and penetrating blue-grey eyes I was unable to forget, even after only one encounter. He spun out an extremely strange train of thought about the structure of the earth and asked me whether I would be willing to help him, a physicist, with geological facts and concepts. As off-putting as this idea seemed to me, the man himself became my friend. As long as we could find time from our military responsibilities, we were able to work together informally. . . .The man was Alfred Wegener.*[17]

In the meantime Wegener had decided to pursue an academic career. He chose the university town of Marburg, small and quiet, yet bustling with scientific activity. At that time Marburg had no professors of meteorology, much less a department, but Wegener found a generous mentor in the experimental physicist Franz Richarz and an extremely stimulating colleague in the geologist Hans Cloos, who was five years his junior. The senior professor of geology in Marburg was Emanuel Kayser,[18] who was widely known for his outstanding textbook. In 1909 Wegener secured a teaching position at the university on the basis of his paper on the kite and balloon ascensions of the "Danmark" expedition. He thus became a *Privatdozent*, a title which today usually implies a secure civil service position. At that time, however, such lecturers lived on their own resources (of which Wegener had few) supplemented by the small sums their few students paid to attend their courses, by honoraria, and by lecture fees. It was an uncertain existence, which demanded much individual initiative. It was not until 1915 that "Captain" Wegener, age thirty-four, became a professor in Marburg University's Institute for the Physical Sciences (*Physikalische Institut*) at a salary of 1500 marks. A small observatory was also affiliated with the institute; its tower-like structure can

Alfred Wegener in 1910.

The former Institute for the Physical Sciences in Marburg and its observatory. Photographed in 1978.

still be seen today on an old building at Renthof 6 near the present-day institute.

Old catalogs of the university list the courses Wegener taught, beginning with the winter semester of 1909–1910: "Astronomic-geographic position-finding for explorers, with exercises at a time to be announced later, Wednesday, 5 o'-clock." In later semesters he lectured on "The Physics of the Atmosphere," "Atmospheric Optics," and "General Astronomy." The listings from winter 1912–1913 to winter 1913–1914 have only the notation: "Dr. Wegener on leave." During that period he was in Greenland for the second time, now accompanying the Dane Johan Peter Koch, with whom he successfully accomplished a major traverse of the interior ice sheet from east to west in 1912–1913. He appears for the first time as "Prof. Wegener" in the catalog listings for summer 1917. His last catalog entry is for the fall intersession of 1919. We can safely assume that the number of students attending his courses at this small university must have been very low. One of them was Johannes Georgi,[19] who would accompany him to Greenland in 1929 and 1930 (See Georgi's reminiscences in Chapter 9). Beginning in the winter of 1914–1915, Hans Cloos (later to become a well-known German geologist) was also in Marburg, lecturing on geology as a *Privatdozent.*

The years in Marburg up until 1914 and the outbreak of war can be seen as the most productive of Wegener's life. Although his university position offered no financial security, he had confidence in his abilities and capacity for hard work. With the second Greenland expedition successfully completed, he was occupied with many new ideas and plans: his first book *Thermodynamik der Atmosphäre (Thermodynamics of the Atmosphere)* appeared in 1911. Most importantly, his hypothesis of "continental drift" took shape during this period and was published for the first time in 1912. We will discuss this, his major work, in detail later. "I think we can safely say that everything Wegener accomplished afterwards and many of his original ideas had their roots in that period," Hans Benndorf wrote in 1931.

Before her marriage to Wegener in 1913 and during her fiancé's second trip to Greenland, Else Köppen had spent eleven months in Kristiania, today's Oslo. She lived at the home of the Norwegian geophysicist and meteorologist Vilhelm Bjerknes[20] (who later taught at the University of Leipzig), gave German lessons to his children, and learned the Scandinavian languages herself, a skill she was later able to use as a translator. She translated, for example, J. P. Koch's account of his expedition

Alfred, Else, and their daughter Hilde in 1916. Wegener is wearing his army officer's uniform.

of 1912–1913 from Danish to German. Entitled *Durch die Weisse Wüste* (*Across the White Wilderness*), the book was edited by Alfred Wegener and published in Berlin in 1919. She also translated two of Vilhelm Bjerknes' books from Norwegian to German: a biography of Neils Henrik Abel (Berlin, 1930), the gifted mathematician who had died at an early age, and another of Karl Anton Bjerknes, the author's father.

Alfred Wegener's Marburg idyll was abruptly ended by World War I. Wegener, a reserve officer, was called up immediately, and was twice wounded on the Western Front. During a lengthy period of sick leave he was able to complete a detailed depiction of the "original continent"—Pangaea—and the drifting land masses. This work, *Die Entstehung der Kontinente und Ozeane (The Origin of Continents and Oceans)*, was first published in 1915. Wegener was then assigned to the military weather service, eventually seeing duty in Bulgaria and finally in Dorpat, Estonia, a city famous for its German university. The research of Johannes Letzmann of Dorpat on waterspouts (published in Dorpat in 1923) clearly shows the influence of Wegener's work. During those years, Wegener also had other opportunities for doing research. Important work on the meteor of 1916, whirlwinds and waterspouts, and the lunar craters was completed or at least begun during his period of military service.

Hamburg (1919–1924)

> *A source of great pleasure were the informal invitations to the Köppen-Wegener home. Our houses were only a few yards apart, and our children used to play in each other's gardens. . . . I was very impressed, however, on our occasional visits by Wegener's taking his leave of the little company, to attend to urgent work in his study, usually correcting proofs. Without such economy of his time his astonishing productivity would not have been possible.*
>
> —JOHANNES GEORGI[21]

A happy turn of events, amid the unhappy confusion of the end of World War I, brought Wegener the offer of the position of Department Head of Theoretical Meteorology in the weather service at the German Marine Observatory in Hamburg.[22] He was to succeed Wladimir Köppen, his father-in-law, who had retired at the age of seventy-two. Kurt Wegener became a department head there as well, and the brothers were once again united for the next five years. The chances of a university professorship were extremely small at that time, but the University of Hamburg was founded in 1919, and Wegener transferred his "habilitation", the right to hold academic lectures, from Marburg to Hamburg. The geophysical colloquium which he founded, along with the Köppen-Wegener home on Violastraße (now Köppenstraße), became an intellectual focus for Hamburg meteorologists, among whom was Johannes Georgi.

In addition to his book on the lunar craters, *Die Entstehung der Mondkrater*, and many shorter papers on meteorology, the second and third editions of *The Origin of Continents and Oceans* were completed in Hamburg, along with an important new work, *Die Klimate der Geologischen Vorzeit, (Climates of the Geological Past, 1924)*, written together with Wladimir Köppen. On an ocean voyage to Cuba and Mexico in 1922, Wegener and Dr. Erich Kuhlbrodt took measurements of upper air currents over the Atlantic in anticipation of transatlantic air travel. Thus the Hamburg years were also rich in scientific accomplishments.

Graz (1924–1930)

When one comes out of the Alps into the magnificent setting of Graz, it is as if one were coming out of war into peace. The bluff of the Schlossberg towers over the city like a protector.[23]

Wegener's life took an important turn in 1924, when he accepted an appointment as professor of meterology and geophysics at the University of Graz, where he succeeded Heinrich von Ficker.[24] Wegener, a Prussian who had lived most of his life in the north German lowlands, now moved to the Alps and the most southeasterly university of the German-speaking area and became an Austrian citizen. The house at Blumengasse 9 (today called Wegenergasse) now became the home not only of the Wegener family but also of the Köppens, who also de-

This tablet was placed on the wall of Wegener's former house in Graz in 1980. The inscription reads, "Alfred Wegener, professor of meteorology and geophysics at the University of Graz, lived in this house from 1924 to 1930. His theory of continental drift initiated a development in the earth sciences which produced a revolutionary change in our picture of the earth."

cided to give up Hamburg for Graz, in spite of their advanced age.

In her biography of her husband, Else Wegener entitled the chapter on the period 1924 to 1930 "The Happy Years in Graz." As far as Wegener's scholarly achievements are concerned, that title is not particularly apt; the Marburg years would fit it better. But otherwise, her choice of phrase is easy to understand: a happy family life with his wife and three daughters; the presence of his father-in-law, who at 78 was still a more-than-able intellectual partner; the pleasant atmosphere of the charming city on the Mur River; and finally, the sense of having achieved the position in life he had aspired to—something that the choice of an academic career had certainly not guaranteed—these all came together here.

Wegener also enjoyed a particularly good relationship with the physicists at the university. The offices of the Department of Geophysics and Meteorology (listed in the university catalog as the "Meteorologic Institute and Meteorologic Station at the Institute for Physical Sciences") were located in the Physics building. Among the physicists, beginning with the eldest, were the theoretician Michael Radakowic, the experimental physicist Hans Benndorf, and, the only one who was a bit younger than Wegener, Victor Franz Hess,[25] a future Nobel laureate who was studying cosmic radiation. The astronomer Karl Hillebrand should also be included here. All of them took an avid interest in geophysical problems. As mentioned above, Benndorf wrote a wonderful, warm, memorial essay for the colleague who was so unexpectedly taken from their circle.[26] On the other hand, Wegener had surprisingly little contact with the geologists, as had been the case in Hamburg as well. We will return briefly to that point later.

On paper, the Graz period lasted six years, but it was actually substantially shorter because of two journeys to Greenland, a preparatory trip in 1929 and a major expedition in 1930. For that reason Wegener's course offerings at the university were not very extensive. According to the catalogs, he began teaching in the winter semester of 1924–1925 with "Optics of the Atmosphere" (2 hours). The last course he taught, in the winter semester of 1929–1930 was "General Meteorology" (4 hours). Considering the small size of his department, it is understandable that Wegener had very few students of his own.

Wegener's years in Graz were taken up mainly with work on continental drift and, as indicated above, with preparing for and carrying out two expeditions to Greenland. These built on his earlier trips to the Arctic and represented the fulfillment of his long-standing desire to lead his own major Greenland venture. Johannes Georgi, whom Wegener knew from Marburg and Hamburg, was also planning to carry out meteorologic experiments in Greenland, and these were now included in the plans for Wegener's expedition. The Göttingen geologist Wilhelm Meinardus also gave strong support to Wegener's plans. The entire undertaking was financed by the *Notgemeinschaft der Deutschen Wissenschaft (Research Aid Society)*, the forerunner of the present-day *Deutsche Forschungsgemeinschaft*. The *Notgemeinschaft* was founded in 1920 with the backing of German universities, technical colleges, and learned societies, to encourage scientific research. On April 1, 1930, Else Wegener said goodbye to her husband and the other expedition members in Copenhagen. On December 13 she received his last letter from Greenland, dated September 20. In November, just a few days after his fiftieth birthday, he died while crossing the interior ice sheet. But it was only in May 1931, after months of anxious waiting and fading hope, that his fate was known with certainty.

3

THE EXPEDITIONS TO GREENLAND

Here you are, sitting in the circle of your
friends,
But you are powerfully drawn to a distant
land.
Hot desert sands and the eternal ice of misty
seas
Are already enticing you away.

Bon voyage. . . .At the threshhold of youth
You'll easily board the ship.
Then, fortune will grant you the sea and wave
And storm and wind that drank your soul.[1]

The Expeditions to Greenland

Throughout his adult life, Wegener was always on the go, off to distant parts. One can well imagine that a scientist attracted to astronomy, meteorology, and geology might succumb to the lure of polar ice. In the following, Wegener's early expeditions are treated only briefly, but the last, ill-fated expedition of 1930 is described in some detail.

Wegener's first three expeditions

The tragic death of Erichsen and his companions is reminiscent in many respects of Wegener's own end.[2]

Alfred Wegener's first exploration of Greenland was as a member of the "Danmark" expedition which spent two years, from 1906 to 1908, on the island. This expedition, which was led by the Dane Ludvig Mylius-Erichsen,[3] consisted of twenty-eight men, including Johan Peter Koch,[4] whom Wegener accompanied to Greenland again in 1912–1913. Mylius-Erichsen was mainly interested in studying the Eskimos of Greenland, but the expedition members also planned to carry out a thorough exploration of the largely unmapped northeast coast between 75 degrees north and Peary Land on the north coast. The group spent the winters of 1906–1907 and 1907–1908 in Danmarkshavn, in a small "winter house," and went out from there by sled. Tragically, in the summer of 1907 Mylius-Erichsen and two of his companions lost their lives when they ran out of food during one such trip away from the base camp. Wegener

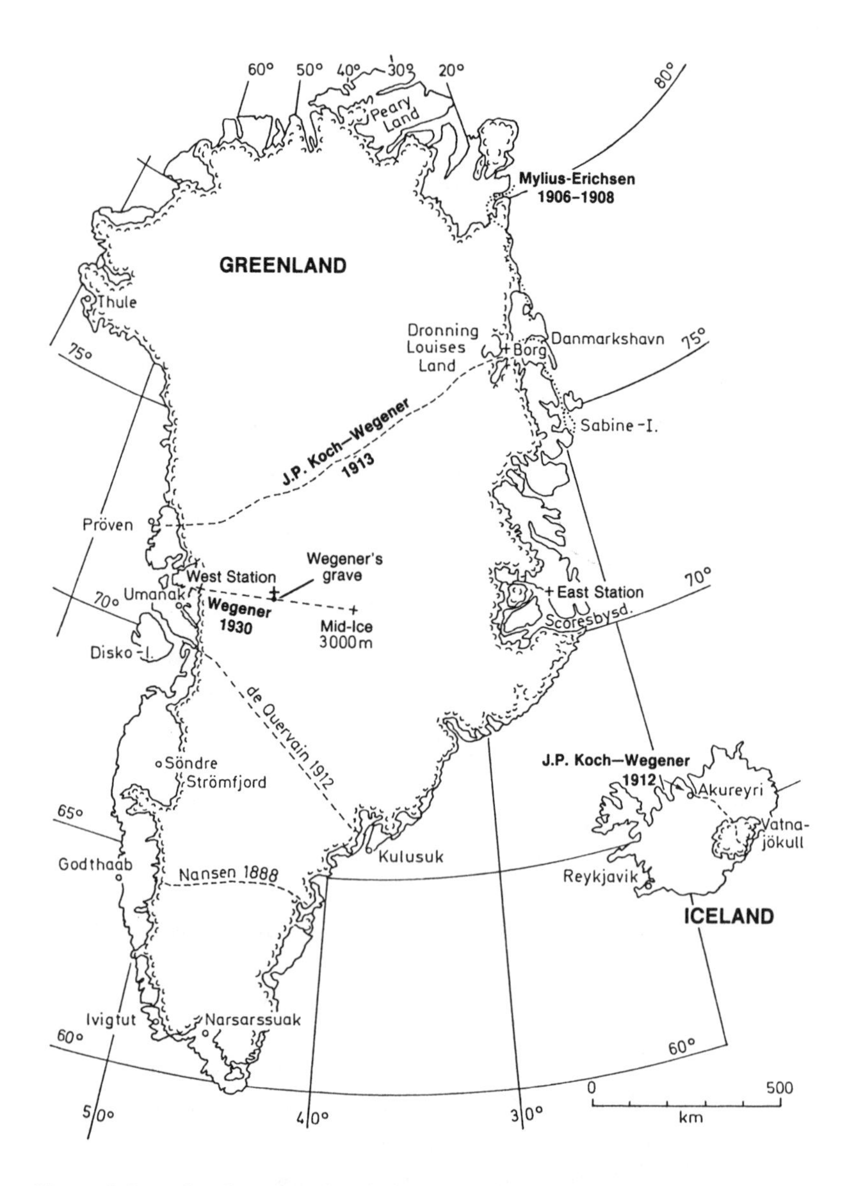

Map of Greenland and Iceland showing the routes of the expeditions in which Wegener participated (marked in bold-face) and of two earlier expeditions by other explorers.

and the other members of the expedition remained for one more year to complete their task.

Wegener's meteorological experiments mainly involved sending up kites and captive balloons to heights of up to 3,000 meters, the first such experiments to be done in a polar climate. His data were incorporated into the paper which gained him his first teaching position in Marburg (Wegener, 1909). Participation in this expedition was also extremely worthwhile for him as it gave him the chance to become thoroughly familiar with living and travelling in Greenland. In addition, he became acquainted with Johan Peter Koch, the leader of the next Greenland expedition.

Wegener's second expedition to Greenland took place four years later, from 1912 to 1913. This expedition was carried out by only four men: Johann Peter Koch, the forty-two-year-old leader; Alfred Wegener, the expedition's scientist; Vigfus Sigurdsson of Iceland; and Lars Larsen of Denmark. In the summer of 1912, the steamship "Godthaab" brought them first to Iceland, where they stayed for two and a half weeks (see Chapter 4) and then to Danmarkshavn, which had also served as base camp for the "Danmark" expedition. Since Mylius-Erichsen's plans to establish a winter camp on the ice sheet for the "Danmark" expedition had failed because the dogs could not manage the steep ice slopes at the edge of the glacier, Koch had brought sixteen ponies from Iceland for this expedition. Using these ponies, the group was able to set up a winter camp at Borg, between Dove Bay and Dronning-Louise Land, thus making overwintering on the ice sheet possible for the first time. At the end of April 1913, they began the great trek across the ice cap at its widest point, climbing to heights of up to 3,000 meters. After much hardship and when they were almost out of food, they were found by some Eskimos who helped them reach the small settlement at Pröven on July 15. None of their ponies survived this trip, although they tried to save the last

Wegener in 1913 in the winter camp at Borg, Greenland.

one by pulling it on one of their sleds—a remarkable demonstration of friendship to a faithful animal companion, considering their own desperate situation.

This may well have been the expedition that Wegener found the most personally satisfying. Although it was marked by deprivation and risk, in the end it was completed successfully with no loss of life. His earnings from the expedition amounted to 15,100 marks, the greater part paid by the German government and the remainder by a few private individuals.

Wegener did not return to Greenland until the summer of 1929, when he led a short expedition made up of only four men.[5] The goals of this expedition, in which Johannes Georgi, Fritz Loewe,[6] and Ernst Sorge[7] joined Wegener, were to find the best site on the west coast to ascend to the interior ice sheet and to try out several new techniques that were to be used during the main expedition the following year, such as measurements of the thickness of the ice. They finally chose Umanak Bay as the site of their base camp, north of Disko Bay and 200 kilometers southeast of Pröven, where Wegener and Koch had ended their traversal of the ice sheet in 1913. The *nunatak* (in the Eskimo language, *nunatak* means an ice-free outcropping of rock) called "Scheideck", above the small Kamarujuk Fjord, was chosen to be the site of the "West Station" for the next expedition.

The major expedition led by Alfred Wegener[8]

Somebody handed me a telescope. It was now almost twenty years since the site had been used. "Mid-Ice" would have to be about fifteen meters beneath our feet. After so many years, nothing could possibly remain of the few installations that had been above ground. And yet, when I took the telescope and began to scan the surrounding area, it didn't seem to me or to anyone else that this was absurd, although, of course, it was useless. No one spoke. My heart pounding, I scanned the horizon. When I lowered the telescope, the silence was broken: Well?. . . .Nothing?", someone asked. "No," I answered. "Nothing. Nothing at all."[9]

Blasting on the ice cap using 73 kilograms of TNT to produce an "artificial earthquake" in order to measure the thickness of the ice.

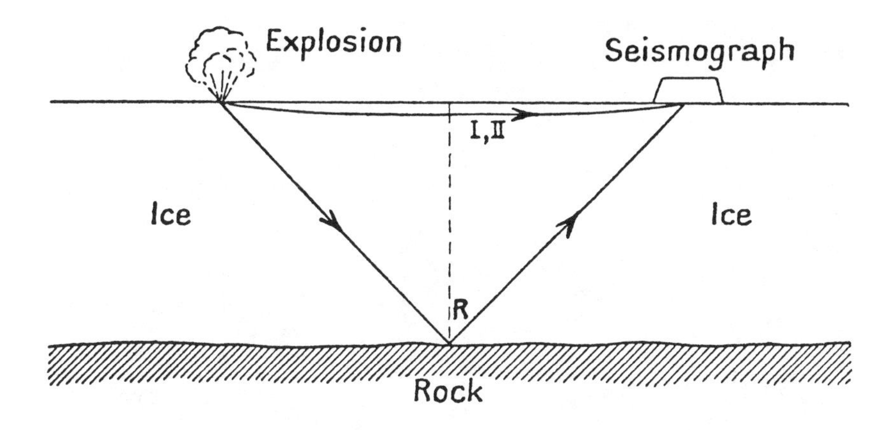

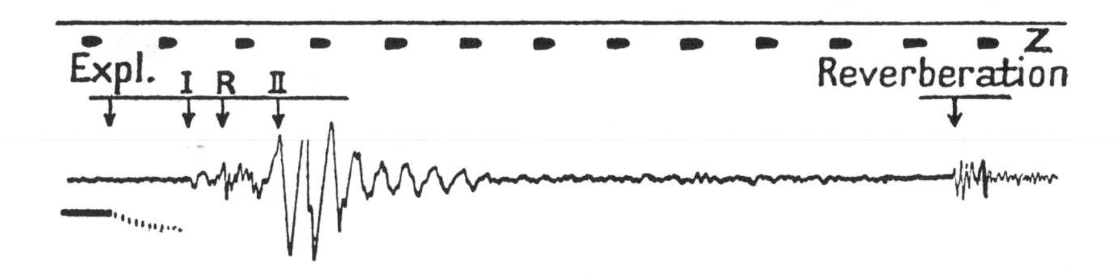

A diagram from a report of the Wegener expedition showing how explosions were used to measure the thickness of the ice. Above: Principle of seismic ice-thickness measurement. Below: A recording strip.

The expedition of 1930 to 1931, planned and led by Wegener, numbered twenty men and was the most ambitious of the four Greenland ventures in which he was involved. Wegener established three camps along the seventy-first parallel, so that meteorological and geophysical research could be carried on continuously across the entire ice sheet. The base camp was set up on the west coast of Greenland ("West Station" at an altitude of 975 meters), and from there the second camp, "Mid-Ice," was established and supplied. "Mid-Ice" was 400 kilometers from the West Station, at an altitude of 3,000 meters. The third

Preparing to pull a motorized sled up onto the Greenland ice cap using humans and Icelandic ponies.

camp, independent of the others, was "East Station" on Scoresbysund on the east coast.

Wegener had ordered two propeller-driven motorized sleds to be built in Finland, especially for the expedition. He hoped these sleds would make faster and more certain progress than dog sleds, although he also used dog sleds, as well as Icelandic ponies. His high hopes for the motorized sleds, however, were only partially fulfilled as the sleds were unable to climb the steep Greenland slopes. Among the new techniques employed successfully in the Arctic for the first time by this expedition was the use of explosives to measure the thickness of the ice sheet using seismic methods.

This fourth expedition of Wegener's, which should have represented the high point of his research in Greenland, was

Sled stuck on a steep slope on the glacier.

plagued by bad luck from the beginning. Their landing in Kamarujuk was thirty-eight days late, on June 16, due to unusually unfavorable ice conditions. "Those thirty-eight days decided the fate of the expedition for 1930," Kurt Wegener wrote.[10] As a result of the tremendous difficulties which they experienced in moving their equipment, including the motorized sleds, up onto the ice sheet, the West Station could not be put into full operation until much later than expected. By that time, with winter setting in, it was too late to use the motorized sleds, and that, in turn, meant that the "Mid-Ice" station, where Georgi and Sorge would be spending the winter, could not be as well supplied as had been planned.

Wegener soon realized that his expedition was heading toward a difficult, and, ultimately, disastrous situation. As early as May 5, 1930, he wrote in his journal,[11] "We must lower our expectations yet another notch." On June 9, their thirty-first day of waiting to land, he wrote, "The day is gloomy and my spirits even more so. . . .Our expedition's program is slowly being seriously jeopardized by the obstinacy of the ice." On July 18–19: "Today I am truly depressed. How are we ever going to get on with things?" On August 5: "It seems that the chance of success is slipping away from us, because, on balance, our vehicles aren't working efficiently enough." On August 6: "August will see the deciding battle, and I have to confess that I am worried when I think about it, particularly with respect to the central station out on the glacier. Before long, I will be out of ideas. . . .I'm afraid, really afraid, that we're not going to make it."

Another journal entry in the same tone of resignation refers to a plan discussed in Graz of having his wife join him in Greenland the next summer:

> I don't think Else should come next year after all. Even if she preferred to live on the boat,[12] it would be a terribly uncomfortable life for her. We would never, or almost never, be able to be alone, we'd get lice.—No, when I dream, I actually dream of something completely different—the Adriatic, and vacation trips with no mountain climbing or other semi-polar adventures. What makes it easy for me to tolerate the many unpleasant aspects of daily life here is the great mission that must be completed.

Finally, in a similar vein, he wrote to his brother on the same day:

> In short, life here has its dark side. A person wouldn't be able to put up with most of it, if he didn't know that after a certain number of months are marked off on the calendar, he can go back home and live as he sees fit. And then, thank goodness, the obligation to be a hero ends, too. . . .Even a paradise eventually loses its ability to make a person happy. I can see the time coming when it will be like that for me with Greenland.[13]

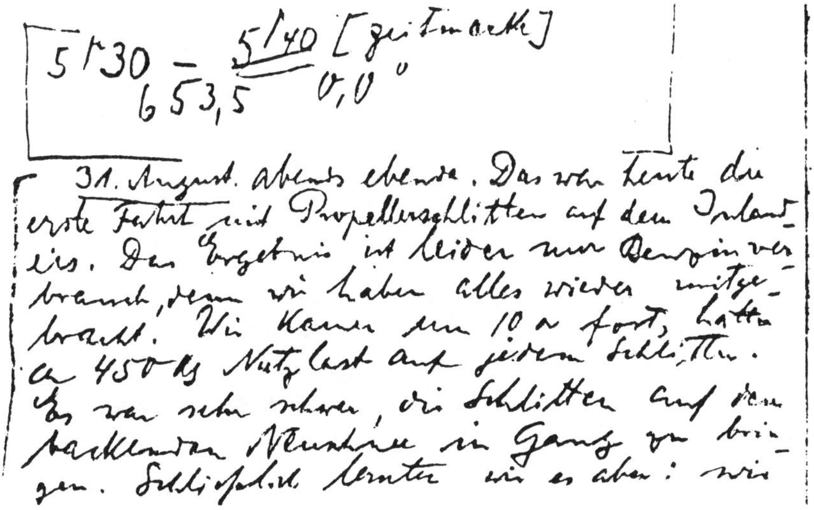
5h30 – 5h40
653,5 0,0°

31. August abends ebenda. Das war heute die erste Fahrt mit Propellerschlitten auf dem Inlandeis. Das Ergebnis ist leider nur Benzin verbraucht, denn wir haben alles wieder mitgebracht. Wir kamen um 10 a fort, hatten ca 450 kg Nutzlast auf jedem Schlitten. Es war sehr schwer, die Schlitten auf dem backenden Neuschnee in Gang zu bringen. Schließlich lernten wir es aber: wir

Excerpt from Wegener's journal about the first, unsuccessful attempt to use the motorized sleds when they were fully loaded on August 31, 1930. At the top of the figure: barometric pressure and temperature (0° C) at 5:30 p.m. Below: "Today we took the motorized sleds out on the ice sheet for the first time. Unfortunately, the result was only a waste of fuel, since we brought everything back again. We set out at 10:00 a.m. with a full load of 450 kilograms on each sled. It was very hard to start the sleds in the fresh, wet snow. But finally we learned: we . . ."

That was no longer the young Wegener!

On September 21, Wegener himself finally set out from the West Station, by dog sled, on a perilous journey to "Mid-Ice," accompanied, for the last 250 kilometers, only by Fritz Loewe and the Greenlander Rasmus Villumsen. In spite of the unfavorable conditions so late in the year, with temperatures as low as −50°C, Wegener was determined to bring supplies to "Mid-Ice" and thus insure the continued operation of the winter camp, which he saw as a vital link in the expedition. Johannes Georgi and Ernest Sorge, the two men at "Mid-Ice," had told him that they would return to the West Station if supplies did not reach them by October 20. Wegener knew very well that it was "unfeasible" for the two of them to try to walk back (without dog sleds); they would "freeze to death en route." On his way to "Mid-Ice" Wegener wrote in one of his last letters (dated September 28), "It's a matter of life and death." Wegener's trip by dog sled was disastrous, primarily due to the extreme cold, wind, and snow. The trip took forty days, rather than the anticipated twenty days. The three men arrived at "Mid-Ice" without any supplies and with Loewe suffering from severe frost-bite. At "Mid-Ice," Georgi and Sorge had, in the meantime, realized that they could not safely return to the West Station and they had decided that if they were very careful they would be able to eke out their supplies for the duration of the winter.

Georgi described the extremely primitive conditions in the "ice cave" in an unpublished letter to friends in Hamburg; dated November 28, 1930, the letter did not leave "Mid-Ice" until May of the following year:[14]

> Your letter arrived along with several others at the beginning of the month, brought by Wegener himself after an adventure-filled sled trip. They started out from the West Station with fifteen fully-loaded sleds and ended here with three empty ones. Because of unexpectedly bad weather, almost all of the Eskimo members turned back, and only Wegener, Loewe, and the Eskimo Rasmus continued, arriving here—without supplies or fuel—on October 30 with the temperature at −54°C, a major achievement in the history of Greenland exploration, considering the weather conditions and that it was so late in the year.
>
> At the beginning of October, when the temperature in our tent was going down to −35°C at night, the two of us, Sorge and I, had taken refuge from the cold by digging down three meters into the snow. We had been thinking of this "ice cave" as only temporary quarters. But now it must remain our home until next summer, because when the fifteen-sled convoy fell apart, our win-

ter house, which we had built with such care in Hamburg, got "left by the wayside" along with all kinds of other more or less vital necessities. So, overwintering here will be much more primitive, and thus more taxing both physically and mentally, than we had thought and than is usually the case.

But still, the lack of petroleum, which forces us to work—or to vegetate in our sleeping bags—at temperatures of about −10°C, is easier to bear, than the fact that on the sled trip from the West Station my comrade Loewe froze both of his feet, and, since a return trip with Wegener and Rasmus would have meant risking all of their lives, he had to be left behind here in that condition. No knowledge of medicine, no book with instructions for this kind of problem, no surgical instruments, no anesthetic, a minimal supply of bandages and disinfectant—that was our desperate situation. The results so far: amputation of all his toes (except the fourth and fifth on the left) with a pocket knife, in the hope that the necrosis won't spread to his feet, or even further, in spite of the horrible wounds and our less than adequate resources.

Thus, we must spend the winter here under considerably more difficult circumstances than we could have foreseen. On top of that is the torment of not knowing whether Wegener and Rasmus made it back to the coast or perhaps fell victim to the merciless cold on the way. Nor do we have a radio—that too got left behind somewhere along the way. So we are, in fact, completely cut off from the world until the first sled trip of 1931, probably in May or June, i.e., seven or eight months from now. Of course, we don't doubt for a moment that we'll survive that long even under these conditions. It will just be a little different than we had thought.

Loewe, whose injuries gradually healed, survived the long Arctic winter in the "ice cave." Since there was barely enough supplies at "Mid-Ice" for two people, and since Loewe was not able to travel, it was decided that Wegener and Villumsen would leave, despite the obvious dangers of the return trip to the West Station. Wegener and Villumsen, after only one day of rest, but physically "in excellent shape," started back with seventeen dogs on Wegener's fiftieth birthday. Both perished on that journey, Wegener probably dying in mid-November and Villumsen shortly thereafter.

For almost seven months, however, the fate of the two men was unknown, both at the West Station and at "Mid-Ice", since there was no radio contact between them, and rescue operations were impossible during the winter. Not until May 8, 1931, were Wegener's remains found, 189 kilometers from the West Station, i.e., approximately halfway between the two camps. The burial site had been marked by Villumsen with Wegener's skis.

> Wegener's body was found sewn into two sleeping bag covers. He lay on a sleeping bag and a reindeer skin, three-quarters of a meter beneath the snow level of November 1930. His eyes were open, his expression relaxed, peaceful, almost smiling. His face was rather pale and looked more youthful than it had before. His nose and hands showed evidence of frostbite, as is common on such trips.

> . . . [All of the evidence] indicates that Wegener did not freeze to death while underway, but rather that he died lying down in his tent, probably of heart failure brought on by overexertion. When Wegener died, Rasmus must still have been alert and in good condition. It was touching to see the care with which he had buried Wegener; and one had to admire the pains he had taken to dig and mark the grave. Evidently he took Wegener's pack, which would have contained the journal of this last trip, along with him, intending to bring it back to the West Station.[15]

How and where Villumsen died was never determined, although the remaining men searched for his body after Wegener's grave was found. This courageous man was only twenty-two years old at the time of his death.

Greenland's ice has remained the site of Wegener's grave. A cross six meters high, made of iron rods, and erected by his

The last photograph of Alfred Wegener, together with his companion Rasmus Villumsen, taken on November 1, 1930, at "Mid-Ice" before they left on their ill-fated trip back to the West Station. The temperature was approximately −50° C.

Alfred Wegener's grave on the ice cap of Greenland, photographed in 1931. The burial site was marked with an iron cross, six meters high, by the expedition members.

comrades in August 1931, marked the site for a time. But like all other traces of the expedition which were to be seen on the ice sheet in 1930–1931, Wegener's grave has long been buried deep in ice and snow.

When news of Alfred Wegener's death reached Germany in 1931, his brother Kurt departed for Greenland in order to take over the leadership of the expedition and bring it to completion (for Georgi's account of this expedition, see Chapter 9).

A question of guilt?

Was Wegener's tragic death unavoidable, or can one or another of the expedition members also be held responsible, i.e., is

Kurt Wegener at the West Station in 1931.

there a "question of guilt"? One of Wegener's journal entries (September 3, 1930), which referred to the inadequate supplies of "Mid-Ice," was often cited in this connection, especially in relation to Johannes Georgi: "I'm afraid it will come back to haunt me that I gave in to Georgi's pressure and left it up to him." But there can be no doubt that as the leader of the expedition, Wegener was aware of the precarious situation at "Mid-Ice" from the beginning. Therefore, Georgi can not be blamed for it. Both men, Wegener as well as Georgi, had counted

"Mid-Ice," the research station on Greenland's ice cap. The entrance to the underground station is to the left of center. The tower was made of blocks of ice. The wooden structure on the right contained weather instruments.

on the motorized sleds to bring the necessary food supplies and fuel to "Mid-Ice" in time. When the sleds broke down, primarily due to adverse weather conditions, but also because their engines were not powerful enough, all of Wegener's plans were upset. The journal entry of September 3, the only one of its kind, can only be explained by Wegener's mood of despair at that time. Even Georgi's and Sorge's ultimatum that they would return to the West Station, if supplies hadn't reached them by October 20, had no deciding influence on Wegener's last trip. For Wegener, the idea of giving up "Mid-Ice" was out of the question. The year-round operation of the station was the raison d'etre of the entire expedition; the *Notgemeinschaft* had given him half a million marks for the trip, an enormous sum in that time of financial crisis. Wegener began the return

A diagram of the underground layout of "Mid-Ice". T, theodolite (a surveying instrument); L, living room; Ba, barometer room; B, balloon-filling room; G, gas-production room; St, store room; S, shaft.

trip from "Mid-Ice" on November 1, 1930, with the comforting knowledge that Georgi and Sorge were remaining at their post.

Tragically, Wegener's journals of his last trip to "Mid-Ice" were lost, and what is especially tragic is that the loss was due to the concern and loyalty of his companion Villumsen. He presumably took the journals on the entirely reasonable assumption that they were of the greatest importance and needed to be brought to safety. Villumsen could not have foreseen that he would disappear forever. One can assume that during those weeks Wegener had been grappling once again with the problems that beset his expedition, and that he had recorded those thoughts in his journals. During his two days at "Mid-Ice", "he wrote in his diary for hours," Ernst Sorge reported. Those irreplacable documents would doubtless answer many questions.

In conclusion, it should be pointed out that the President of the *Notgemeinschaft*, Dr. Friedrich Schmidt-Ott, declared in a public statement in the fall of 1931 that there was no question of guilt as far as Wegener's death was concerned.[16]

The Greenland expeditions in retrospect

> *His extraordinary calmness, his willingness to make sacrifices for the sake of his work, his affability, and his sense of fairness made him an ideal expedition leader. And yet, a number of his friends might have regretted that such an excellent and disciplined mind could not have been dedicated solely to academics instead of being drawn so strongly to distant shores.*
>
> —WLADIMIR KÖPPEN[17]

Wegener's expeditions to Greenland are among the greatest pioneering ventures in the history of the earth's exploration. They took him to inhospitable places that were completely, or almost completely, unknown; places that were literally "empty spaces" on the map. They were carried out before modern technology made it possible to overcome most of the difficulties of such expeditions and make them practically risk-free. For example, one can contrast the 400,000-kilometer flights to the moon with Wegener's last, ill-fated trip back to the West Station, a distance of only 400 kilometers, which had taken forty days to cover on the way out.[18] Probably the personal involvement of the expedition's leader in a trip to "Mid-Ice" would not even have been necessary if there had been radio contact between the two camps. The West and East Stations both had

A motorized sled on the ice cap. The sleds had been built especially for the expedition by the Finnish government airplane factory; they had a 110-horsepower airplane engine mounted on the rear, ran on hickory wood skids, and were painted red.

radios, and "Mid-Ice" was supposed to have been radio-equipped as well. But "the weather conditions on the ice sheet in the fall of 1930 unfortunately prevented the radios from reaching their destination" (Walter Kopp, 1933).

Overall, however, many more technical resources were available to Wegener, especially in the later expeditions, than had been available to earlier Arctic explorers, and in Wegener's own opinion, the motorized sleds represented a major step forward.[19]

> Horses on the glacier, dogs and motorized sleds on the ice sheet: that is the right way to go about it. We are beginning a new era in Arctic exploration. We have to start from scratch, as far as all of the measurements we want to take, and are able to take; and

The numerous crevasses made travel difficult for both human and dog.

what we do here will form the basis and set directions for future Arctic research. How wonderful that we are the ones permitted to do it, to take these pioneering, even—considering the many airplane crashes in the Arctic—life-saving steps. (Journal entry, August 29, 1930.)

Here Wegener was mistaken, and a few days later he had already backed off a bit: "Yes, the motorized sleds are conquering the white wilderness! It's just that you can't demand more of them than they are able to deliver" (journal entry, September 7, evening). It is interesting that Wegener, always a balloon enthusiast and the brother of a World War I fighter pilot, did not anticipate the imminent development of air travel and its importance, for example, in future Antarctic exploration, although, of course that development was almost unbelievably rapid.

As far as technology and equipment are concerned, one can hardly compare Wegener's last Greenland expedition with today's highly developed Antarctic research. Nevertheless, it belongs in the same class as modern research expeditions in which systematic scientific experiments are of primary importance.

Foreign Minister Friedrich Schmidt-Ott, the President of the *Notgemeinschaft der deutschen Wissenschaft*, summarized the goals of the German Greenland Expedition of 1930–1931 in the preface to the first volume of its *Wissenschaftliche Ergebnisse* (*Scientific Experiences* 1933):

> This expedition was intended to produce a new generation of scientists with arctic experience so that Germany will not be eliminated from international scientific competition in arctic research. Further, geophysical techniques were to be tested that seemed to hold promise for scientific research. Finally, the meteorological conditions on the ice cap were to be investigated, in view of their noticeable influence on European weather patterns and their great importance to plans for air travel between Europe and America; and at the same time, to study the icebergs which are released into the sea from the great glaciers of Greenland as these icebergs represent a substantial danger to sea travel.

During the expedition, a large volume of meteorological, climatological, glaciological, and geophysical data were collected, including those from the first year-round research in the interior, under remarkably difficult conditions. These sets of data still have importance today. With the measurements of the thickness of the ice sheet (they were able to measure to a depth of 1800 meters), the expedition introduced this type of

Alfred Wegener boring a hole in the ice in order to measure the temperature of the ice.

investigation to research in Greenland and the Antarctic. Later expeditions were able to establish that the thickness of the ice sheet in Greenland reaches 3400 meters and in the Antarctic, more than 4000 meters.

Wegener knew that there were great risks involved in arctic exploration from his participation in the Mylius-Erichsen expedition. Unfortunately, he proved this statement with his own tragic death. Benndorf began his memorial essay with a quotation from one of Wegener's notebooks; even if the words did not originate with Wegener, they are certainly appropriate: "No one who wants to accomplish something worthwhile on this earth can expect to accomplish it, except under one condition: I will accomplish it or die." Kurt Wegener's comment in connection with the Spitzbergen rescue operation for the Schröder-Stranz expedition in 1914[20] is in a similar vein: "If no one dared the impossible, there could be no greatness on earth." Of course, those statements seem overly dramatic and, as far as the natural sciences are concerned, they would probably be valid only for explorers.

When one surveys Wegener's work from today's perspective to see what remains valid, or even controversial, fifty years later, it is clear that his work in Greenland is only of secondary importance, even though his expeditions, like others characterized by hardship and adventure, captured the public attention of the time, thanks to the gripping accounts written by Wegener himself and his colleagues. And yet, there is no doubt that Greenland was of vital importance in Wegener's life. From boyhood on he had an almost passionate interest in arctic exploration with its attendant adventures, exertions, and risks. Wegener was a true "Viking of Science," as Heinrich von Ficker stated in a memorial essay.[21]

As early as 1906, Wegener had written in his journal during his first expedition to Greenland: "Out here, there is work worthy of a man; here, life takes on meaning" (December 25).

Alfred Wegener in Greenland.

In a commemorative address the meteorologist August Schmauss quoted from an article published by Wegener in 1908:

> Alexander von Humboldt states somewhere in his writings that there is so little of interest in the polar regions that expeditions there are not worthwhile. If he could have stood as we did, under

> the flickering northern lights, with an overwhelming feeling of insignificance at the sight of this phenomenon of nature, a phenomenon that humans did not discover, but have simply experienced from time immemorial, he would never have said such a thing. Above us the shining curtain unfolded in mysterious movements, a powerful symphony of light played in deepest, most solemn silence above our heads, as if mocking our efforts: Come up here and investigate me! Tell me what I am![22]

But Wegener's true and enduring importance, as already indicated, is not found in his arctic expeditions but in an entirely different sphere. Greenland was only tangential to his major work on continental drift which we will discuss in later chapters.[23]

4

ALFRED WEGENER AND ICELAND

Ride, ride, and hurry across the sand!
The sun is sinking behind Eagle Peak.
Many unclean spirits are on the move
Now that shadows start falling on the glacier.
God guide my horse!
The last stage of the journey will be hard.

—ICELANDIC POEM[1]

Alfred Wegener and Iceland

The two islands with the misleading—one could say transposed—names Greenland and Iceland are not very far apart (see the map on page 32). Thus, Iceland has served as a point of departure for Greenland ever since Erik the Red first sailed to Greenland in 972. Wegener first became acquainted with Iceland at the beginning of the Danish expedition to Greenland in 1912.[2] On the way to Greenland, Johan Peter Koch and Wegener interrupted their journey for two and a half weeks of travel in Iceland. Beginning in the town of Akureyri in the north, they and four other men crossed the island "for practice", from north to south and back again, traversing Iceland's—and Europe's—largest glacier, the Vatnajökull, which until then had remained largely unexplored. This glacier, however, is tiny compared to Greenland's immense ice cap, as the map shows.

The men covered about 600 kilometers, using twenty-six Icelandic ponies. Where snowmobiles are used today, Koch and Wegener crossed the Vatnajökull on horseback, a feat that still deserves our admiration. Koch was thinking of using Icelandic ponies in Greenland instead of the usual sled dogs, and was testing them on this trip in Iceland. He had previously recognized the value of the horses, which were indispensible for transportation in Iceland at that time, when he had done cartographic research earlier in the southern part of the island. Sixteen of the ponies used in the Iceland trek went on to Greenland with the expedition.

En route across Iceland the party climbed the mountain Kverkfjöll with its many fumaroles and saw the Dyngjufjöll

Johan Peter Koch in Iceland in 1912.

mountain with the great caldera of Askja and its deep lake inside the caldera, in which the German geologist Walther von Knebel and a companion had lost their lives five years previously. Koch and Wegener made rapid progress even though the snow conditions in the summer months on the Vatnajökull glacier were unfavorable; and Koch was justified in pointing out at the conclusion of his report that the relative ease with which they had crossed the Icelandic ice cap contradicted the skepticism occasionally voiced against ancient reports of trails across the Vatnajökull.

Accompanying Wegener and Koch in Iceland was the thirty-seven-year-old Icelander Vigfus Sigurdsson, who was familiar

Icelandic ponies carrying building material on the glacier in Greenland, 1930.

with the terrain. He continued on with them to Greenland, where he supervised the horses and, in general, overcame, in an exemplary fashion, the many unforeseen difficulties that beset Koch's Greenland expedition. He spent the entire winter with Koch and Wegener in the far northeast of Greenland and participated in their subsequent traverse of the ice cap.[3]

Unfortunately, the horses did not survive the exhausting trek across Greenland, but they proved to be very useful in transporting the equipment from the coast on to the ice sheet. Wegener made use of that experience during his 1930 expedition, using Icelandic ponies at the base camp on Greenland's west coast. In a journal entry of August 29, 1930, he wrote: "The Icelandic horses are coming through their trial by fire successfully for the first time here. They weren't able to do so on Koch's expedition, because they weren't used in the right way. We are the first to do it correctly." To take care of the horses, Wegener took three Icelanders to Greenland. Vigfus Sigurdsson was one of them, and the written reports of Wegener and his colleages often praised this dependable and versatile worker and his fellow Icelander Jon Jonsson. Sigurdsson and Jonsson went back to Iceland in October of 1930, but Jonsson returned to Greenland in April 1931 with more horses and stayed with the expedition until its conclusion. The third Icelander was Gudmundur Gislason, a medical student; he remained with the expedition at the West Station through the winter of 1930–1931.

Sigurdsson and Jonsson merit attention for other reasons as well. Sigurdsson was proud of his participation in two expeditions, and of being a real "Greenland traveler." On his basalt gravestone in the Fossvogur Cemetary in Reykjavik (he died in 1950 at the age of seventy-five), the epitaph reads only *Graenlandsfari*, "Greenland traveler." Although he was an "uneducated" man, he wrote of his experiences in a book, *Um Thvert Graenland 1912–1913 (Across Greenland)*, published in Reykjavik in 1948.

Jon Jonsson, another member of the 1912 expedition, also wrote about Greenland. During the months of uncertainty about Wegener's fate in the winter of 1930–1931, he published a short paper, "Er prof. Wegener i haettu staddur?" ("Is Professor Wegener in Danger?") He emphasized Wegener's outstanding personal qualities—his ésprit de corps and his readiness to make sacrifices for the sake of the group. To be sure, those very qualities led Jonsson to conclude that the outcome was likely to be tragic.

A touching symbol of the mutual attachment and sincere relationship between the expedition's leader and its members is the name that Jonsson gave his son, born in 1933: Alfred Rasmus Jonsson.

Gravestone of Vigfus Sigurdsson, "Greenland Traveler", who accompanied Alfred Wegener in 1912–1913 and 1930. Fossvogur Cemetery, Reykyavik, Iceland. Photographed in 1978.

Another member of the last two Greenland expeditions who also was familiar with Iceland was Johannes Georgi. In the summers of 1926 and 1927 he worked in the northwest part of the island, measuring the velocity of upper air currents; he was one of the first to observe the jet stream, which has received a great deal of scientific attention since then.

Iceland's fissures and the theory of continental drift

The remarkable fissures in the Icelandic landscape—sometimes whole "swarms" of them—are among the greatest geological sights of the island. But Wegener did not include them in the framework of his hypothesis of continental drift (see Chapter 8), which is actually quite surprising, since the large-scale fissures in the earth's crust played an important role in Wegener's thinking. They are the starting point for the separation of the continental plates and in Iceland, as indicated above, such fissures are obvious. Why did Wegener ignore them? He certainly considered other fault systems: for example, the extensive fracture zones, which have created narrow, open "graben" structures between the Black Forest and the Vosges Mountains in Europe and the elongated, deep basins which form the Dead Sea and the east African lakes, such as lakes Tanganyika and Nyassa. Wegener characterized them as the "preliminary stages of a complete separation" of two sections of the earth's crust, just as he saw the Atlantic Ridge as a very advanced stage of the same process.

However, in his opinion, Iceland, in spite of its wealth of fissures, was not an active fault zone but rather one which became inactive long ago. He saw the island as a "waste product," and that explains its "lack of tectonic interest." Perhaps it is also relevant that he probably never saw the most impressive fissures in Iceland—the famous *Allmannagja* fissure and

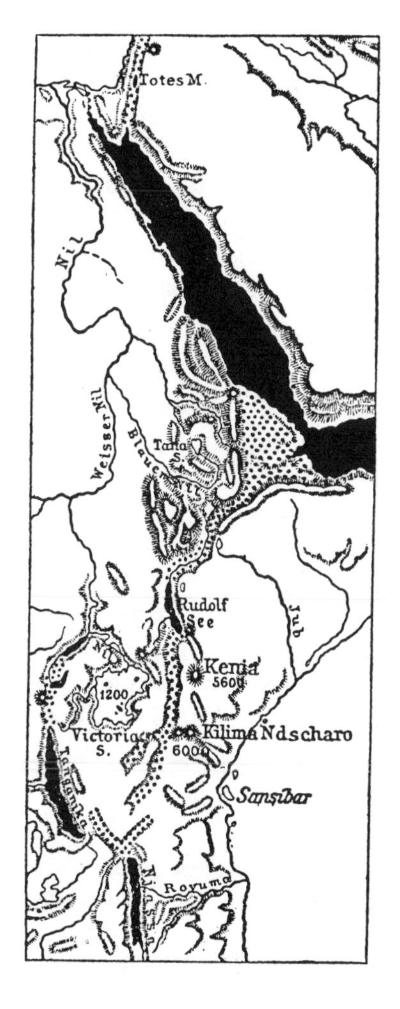

Wegener's drawing of the east African grabens. From Die Entstehung der Kontinente und Ozeane, *Second edition, 1920. The circles indicate graben areas—rift valleys; the black areas are water-filled grabens—lakes.*

its neighboring fissures in the Thingvellir region. (*Almannagja* means "swallows everyone" in Icelandic.)

As early as 1930, however, others began to discuss the possible connection of Iceland with Wegener's hypothesis of continental drift. The first to do so was probably the Dane Niels

Nielsen.[4] In May of 1930 he gave an address entitled "Tectonics and Volcanism in Iceland with Reference to Wegener's Hypothesis" at an arctic colloquium in Greifswald, Germany. He dealt with the same topic in greater detail in a later study of "The Physiography of Iceland" (1934). Here he linked the uniformly oriented tectonic structure of the island, with its "fissure eruptions,"[5] to large-scale stretching of the earth's crust, a view similar to Wegener's theory of continental displacement.

In 1938 a joint German and Icelandic research team under the direction of Otto Niemczyk of the Technical University of Berlin[6] followed in Wegener's footsteps. This expedition took precise geodetic measurements to determine whether horizontal movement, i.e., expansion, can be detected in the Icelandic fissures. Precise measurements were taken at one hundred different points, with the intention to repeat the measurements after a number of years in order to determine whether any displacement of the crust occurred. Niemczyk had previously had a great deal of experience with such measurements. Before he became a professor at the Technical University in Berlin, he had been a mining surveyor for many years in the coal-mining region of Upper Silesia, where he had carried out precise geodetic investigations of depressions in the earth resulting from mining activity.

Ferdinand Bernauer,[7] a member of this expedition who studied the geology of Iceland, wrote: "The Icelandic volcanic zone is a deep scar in the earth's crust, created by the rending of the tertiary basalt layer as a result of rising magma currents." From this he derived the concept of a large-scale circulation of magma taking place far underground, the same concept that Arthur Holmes had already depicted in a schematic profile drawing which is now a part of "plate tectonics." We will discuss this point in more detail in Chapter 7. Thus, Bernauer was one of the first to link the modern explanation of the origin of the mid-Atlantic Ridge to that of the Icelandic fissures.

After World War II, geodetic investigations in Iceland, using more refined methods, were performed by a research group led by Karl Gerke of Braunschweig; and paleomagnetic measurements were carried out by Gustav Angenheister from Munich and his students. It was difficult to find the points in Iceland where the old measurements had been taken. "Perhaps the easily visible signal rods of 1938 were removed during the war."[8] The later measurements provided valuable data but no unequivocal, direct answer to the question of continental drift.[9] In fact, for several years an unmistakable contraction of the fissures was detected, instead of the expected expansion, a result which "came as a surprise to many earth scientists." Obviously, a "certifiably correct interpretation of the complicated movement processes" is not possible at present. But overall, Iceland offers in its countless fissures—both those which have long since been filled by basalt lava and those which are still open—a prime example of a zone of expansion in the earth's crust.

After World War II, geodetic measurements were also carried out by other research groups, as Iceland became a favored sphere of international research activity, especially in connection with the mid-Atlantic ridge, of which it is a part. Thus the island has become much more important to Wegener's successors than it was to Wegener himself. Many supporters of the theory of continental drift have undertaken studies of the geophysics and geology of Iceland, as have their less numerous opponents. Therefore, it is not surprising that the name of the American geologist Arthur A. Meyerhoff appeared in the guest registers of Icelandic hotels in 1972, Meyerhoff being one of the most vociferous of the opponents of Wegener and plate tectonics, along with the Russian geologist Vladimir V. Beloussov. In Beloussov's opinion, it was surprising that the theory of continental drift had ever been taken seriously. The Russian geologists worked in Iceland for several summers and published their results in four large volumes under the title *Iceland and the Mid-Ocean Ridges* in 1977–1978. A new collection of papers, *Geodynamics of Iceland and the North Atlantic area*, edited by the Icelander Leo Krisjansson and including contributions by many researchers outside Iceland, also attests to the central importance of the island to the theory of plate tectonics. Finally, Kristjan Saemundsson provided a wealth of information in his recent survey of the Icelandic fissure swarms (1978).[10]

Thus, although Wegener's Icelandic experiences contributed little to his concept of continental drift, the geology of Iceland has provided some of the strongest evidence for movement of the earth's crust.

5

ALFRED WEGENER'S SCIENTIFIC WORK

A great genius will seldom make his discoveries on the paths of other men.[1]

Alfred Wegener's Scientific Work

The versatile meteorologist

Meteorology, including meteorological optics and kite ascensions, is really the area where I work best and with the most happiness.

—Alfred Wegener[2]

In his professional life, Wegener was above all a meteorologist. As we have seen, meteorology was of primary importance in his expeditions to Greenland; and among his scientific writings, those pertaining to meteorology outnumber the rest. The bibliography of his most important publications in the appendix indicates that emphasis, while the full list of 170 items, compiled by Benndorf in 1931, provides a more complete picture, as well as an idea of the energy Wegener devoted to his research. Obviously, sometimes his family must have had to take second place.

Wegener's most comprehensive publication in the field of meteorology was *Thermodynamics of the Atmosphere (Thermodynamik der Atmosphäre)*, a textbook of 331 pages which was published in 1911. Wegener was only thirty when he wrote it, quite young to be doing so. He had sent the manuscript to the meteorologist at the Marine Observatory, Dr. Wladimir Köppen, for review, even though he knew him only slightly, thereby unintentionally paving the way for meeting Köpppen's daughter, Else. Hans Benndorf found the book "wonderful," as did

August Schmauss,[3] and the respected Russian climatologist Alexander Woeikoff declared that a new star of meteorology had risen.[4] Köppen himself, in the brief obituary note he wrote for his son-in-law years later, pointed out that the book showed Wegener's "special talent for explicating difficult problems simply and clearly with a minimum of mathematics, and yet with no loss of precision." Wegener had no great love for mathematics, which perhaps explains why many students preferred that particular textbook. It went through a second, unrevised edition in 1924 and a third in 1928. Wegener wrote on the same topic in the *Handbook of Physics (Handbuch der Physik*), which was edited by H. Geiger and K. Scheel, and on other aspects of atmospheric physics in Gutenberg's geophysics textbook and Müller-Pouillet's physics text. After Wegener's death, Kurt Wegener oversaw the posthumous publication of his brother's *Lectures on the Physics of the Atmosphere (Vorlesungen über die Physik der Atmosphäre, 1935)*.

Wegener came back again and again to research on tornadoes and waterspouts, and did fundamental work on atmospheric halo effects,[5] all meteorological phenomena which allowed him to proceed from direct observation. However, he was not a "theoretical" meteorologist.

Several times Wegener turned his attention to objects which are of meteorological interest only in the last stages of their existence: meteors. An almost classic study by him in that area was titled "The Impact of a Meteor at 3:30 p.m. on April 3, 1916, in Kurhessen." The fireball and its disintegrations had been observed by many witnesses, and the Institute for the Physical Sciences in Marburg collected their statements. "Captain" Wegener was granted leave to investigate, and he succeeded in narrowing the impact zone of the suspected meteorite to the area around Treysa, thirty kilometers northwest of Marburg. The geologist Emanuel Kayser wrote in his unpublished memoirs[6] that the Society for Scientific Research in Marburg had offered a prize for finding the meteorite:

> But the summer and winter went by without anyone succeeding, and we concluded that it must have landed in the extensive forests in the area around Treysa. And, in fact, that is exactly what had happened. In the spring of 1917, Professor Richarz, the chairman of the Society for Scientific Research, received a report that a forester had discovered the stone in a forest north of Treysa. Richarz set out immediately for the site with me and an aide from the Institute for the Physical Sciences. Unfortunately, the meteorite had already been dug out and taken to a nearby village. However, we searched out the impact site, a small glade, and determined that on impact the stone had driven into the ground at an angle, leaving a cavity over a meter deep. The valuable stone, a flat meteorite, weighing sixty-three kilograms, measuring about a meter across, and consisting, as usual, of metallic iron, was then taken back to Marburg and housed, first in the Physical Sciences Institute, and then in the Mineralogical Institute.

This was probably the first time that a meteorite was found primarily on the basis of systematic calculations.

From meteorites it is not far conceptually to the craters of the moon. The moon's craters, of course, are almost completely outside the provenance of meteorology, but we should mention Wegener's detailed study, *The Origin of the Moon-Craters (Die Entstehung der Mondkrater, 1921)*. Like some of his other scientific writings, this one is still astonishingly up-to-date. In this short book Wegener asked: were the craters on the moon formed by volcanic action, or by the impact of meteorites? Wegener proceeded to answer the question with controlled experiments, using powdered cement. His experiments produced impact craters very similiar in appearance to the craters of the moon, and Wegener thus concluded that most of the lunar craters were formed by the impact of meteorites or other similar bodies. That interpretation is entirely consistent with the one prevailing today—following a series of successful lunar voyages that were undreamed of in Wegener's day.

However, Wegener's experiments never received any attention and have long since been forgotten.[7]

The origin of continental drift

We now come to a completely different area of research, the "science of the structure of the earth's crust and the movements and forces which shaped it," commonly called tectonics by geologists.

By 1910 at the latest, Wegener's attention was drawn to a particular feature on a world map. "Doesn't the east coast of South America fit exactly against the west coast of Africa, as if they had once been joined? The fit is even better if you look at a map of the floor of the Atlantic and compare the edges of the drop-off into the ocean basin rather than the current edges of the continents. This is an idea I'll have to pursue" (from a letter to Else Köppen). As one can see, even in his very first reflections on the question, Wegener did not proceed from the fit of the present-day coastlines but rather, entirely correctly, from the edges of the undersea platforms which are adjacent to the coasts, the continental shelves. That insight, therefore, was not a discovery of later researchers, as has occasionally been assumed. (There is an American study devoted solely to that topic![8])

The remarkable symmetry of the coastlines had, of course, been noticed long before Wegener, beginning 300 years ago with the philosopher and statesman Francis Bacon. Alexander von Humboldt should also be mentioned here. In his book *Kosmos*, he compared the Atlantic Ocean to an enormous valley, whose waters had surged first toward the northeast, then the northwest and finally back toward the northeast. In 1858 in Paris, Antonio Snider-Pellegrini published a book entitled *La création et ses mystéres dévoilés (Creation and its Secrets Unveiled)*, which included a map on which the Americas bordered Africa: a reconstruction of the Carboniferous period was used to explain the similarity of the American and European carboniferous flora (see figure on page 78). The idea seems almost "Wegenerian," if only in that one detail. It was only by chance that this predecessor of Wegener's was rediscovered by an Englishman, Arthur A. Robb. Robb was studying Wegener and his paleogeographic reconstructions in 1930 when he suddenly recalled seeing Snider-Pellegrini's drawing in a book that had been popular during his youth.[9]

Around 1910 three Americans—William H. Pickering, Howard B. Baker, and Frank B. Taylor[10]—were also occupied with the striking shape of the Atlantic, a problem that was obviously in the air at that time. The three shared the notion that the moon somehow played an important role, whether by its separation from or capture by Mother Earth.

In the beginning, Wegener was obviously not aware of these predecessors. Only in later editions of his book did he add "preliminary historical remarks." But these early speculations on the positioning of the continents, as Alexander Du Toit correctly emphasized, are of more interest to historians than to scientists, since they never led to any consistent or enduring theory about the development of the earth. The relationship between Wegener and Taylor, as Anthony Hallam pointed out in 1973[11], is quite reminiscent of that between Darwin and Alfred R. Wallace, who both developed the same idea (the theory of natural selection) simultaneously and independently; Darwin, however, developed it in greater detail and to greater effect.

Nonetheless, Taylor clearly stated in 1910 that Africa and South America had once been joined, had broken apart at the mid-Atlantic ridge (whose importance to this question he recognized), and had moved away from the ridge in opposite directions. Because of this and other similarities to Wegener's depiction, scientists occasionally referred, at first, to the "Tay-

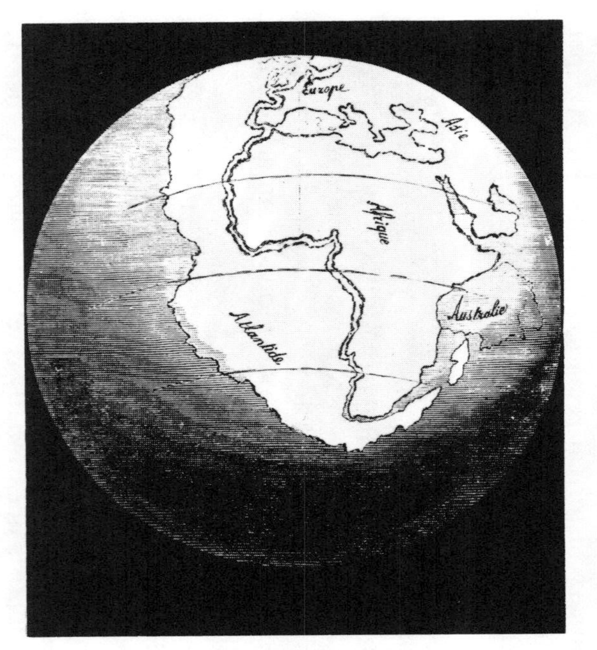

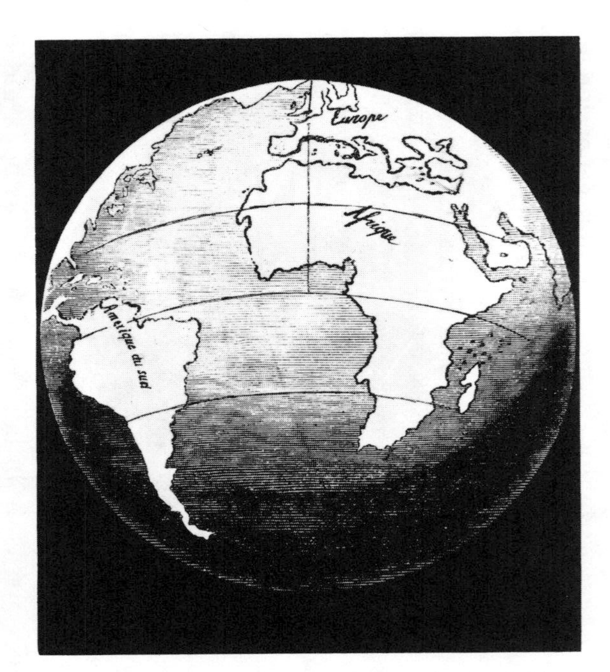

Oldest pictorial representation of continental drift in the Atlantic basin, by Antonio Snider-Pellegrini, 1858. Left: Before separation the continent located in the present-day Atlantic is called "Atlantide"; Right: After separation, "Amérique du sud". Before separation "Australie" is depicted, rather arbitrarily, as lying east of Africa.

lor-Wegener hypothesis." Taylor's hypothesis, however, centered on completely different problems, in particular the origin of the young belt of folded mountains. Wegener was, in fact, alone in developing a comprehensive and consistent hypothesis within a world-wide framework, with the Atlantic coastlines as his point of departure. It was almost exclusively through his efforts that scientific discussion focussed on the idea that the continents have moved independently on the horizontal plane for thousands of kilometers and thus have continuously changed their longitudinal and latitudinal position. In formulating his theory, Wegener drew on a great number of scientific observations, particularly from geology; and he emphasized that a collection of papers he had happened on, which postulated an earlier land bridge between Brazil and Africa on the basis of paleontologic findings, had been an especially important stimulus.

The new hypothesis of continental drift arose from his fortuitous reading of that collection and his chance observation of the congruence of the Atlantic coastlines, thus opening up a completely new chapter in geophysical and geological research. Vertical movements of the earth's crust, isostatic compensation following the advance and retreat of continental glaciation, and the *Nappe* folds[12] of the Alps had long been accepted as matters of fact by geologists. However, the autonomous horizontal displacement of *entire continents* on the magnitude that Wegener postulated was something completely new. In his major work *The Origin of Continents and Oceans* (*Die Entstehung der Kontinente und Ozeane*) which first appeared in 1915, he provided a detailed depiction of that process.

In opposition to the idea of "wandering continents" is the theory that today's land masses have always occupied the same place during the course of the earth's long history, although their formation was affected by the changes caused by the repeated rise and fall of the oceans. Thus, one can speak of "mobilists" versus "fixists" among earth scientists.

Die Entstehung der

Kontinente und Ozeane

Von

Dr. Alfred Wegener

Privatdozent der Meteorologie, prakt. Astronomie und Kosmischen Physik
an der Universität Marburg i. H.

Mit 20 Abbildungen

Braunschweig
Druck und Verlag von Friedr. Vieweg & Sohn
1915

The title page of the first edition of The Origin of Continents and Oceans, *the first, fundamental exposition of Wegener's theory of continental drift, published in 1915.*

Wegener's continental drift hypothesis

Since many earth scientists later misquoted or misunderstood Wegener's hypothesis, it is worthwhile to restate the main points of Wegener's idea of continental drift:

1. The continents and ocean floors are built up of two different types of crustal material: the continents of a somewhat lighter and the ocean floors of a somewhat denser type of rock. Still considered valid today, this differentiation of two major types of rock had been made long before Wegener. Eduard Suess[13] of Vienna, the great geologist of his time, had named the rock types *sal* and *sima*. In the second edition of his book, Wegener followed the suggestion of Georg Pfeffer and used "sial" instead of "sal" (as is now common practice), "in order to avoid confusion with the Latin word for salt." The terms *sial* and *sima* are meant to indicate that silicon-aluminum compounds predominate in the continental sial-rock, as opposed to silicon and magnesium in the sima of the ocean floors. By rock type, sial corresponds most closely to granite (density 2.7) and sima to basalt (3.0).

   Proceeding from these facts, Wegener hypothesized that the continents were gigantic blocks which floated in the denser sima layer much as icebergs float in water. Of course, sial and sima are both solids. Thus, the contention that the land masses move horizontally needs a special explanation, and this is where difficulties—and the potential for opposing arguments—begin to arise.
2. Having assumed the fundamental possibility of drift, what forces are capable of displacing continents? On a map or a globe, it is easy to imagine the process; to find a real mechanism, on the other hand, is difficult. Wegener referred to two types of forces: a) The pole-fleeing force ("pohlflucht"), which, when applied to a rotating sphere, impels

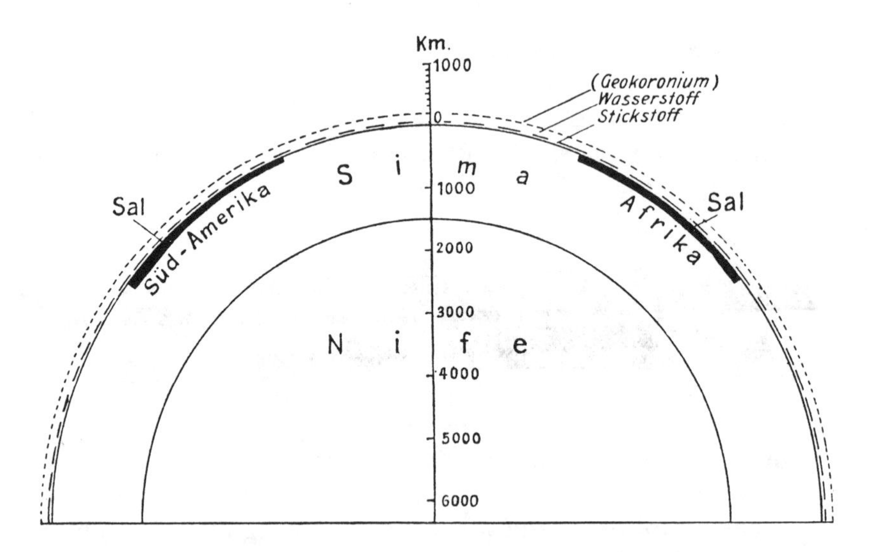

Drawing from Wegener's first paper on continental drift which appeared in the journal* Geologischen Rundschau *in 1912; a cross-section of the earth through South America and Africa, showing the continents— "sal"—floating on the "sima."

the continents toward the equator. That force, then, could explain movements in the direction of the equator. b) The frictional drag of the tides on the solid land masses. This could cause the westward drift of the continents that is especially striking in the case of the Americas, for example.

But Wegener himself saw that these forces were insufficient. In 1915, in the chapter entitled "Possible Causes of Displacement" he wrote that he still considered this question "premature." In the fourth edition of his book, published in 1929, he wrote: "The Newton of the drift theory has not yet appeared."

It is very interesting that in 1929 Wegener first referred, albeit only briefly and cautiously, to the possibility of convection currents in the magma as sources of movement, drawing on the work of Robert Schwinner[14] and G. Kirsch. Today this mechanism is considered the most likely force responsible for crustal displacement.

3. The Atlantic Ocean, which had stimulated Wegener's train of thought in the first place, was cited as the prime example of an enormous fissure which had formed between two separating continents and which had slowly but steadily widened in the course of the earth's history.
4. The great ranges of folded mountains, such as the Alps, the Himalayas, and the American Cordilleras, were created by the crumpling of layers of rock on the leading edges of the drifting continental blocks, in other words, a kind of bow wave.
5. In the Paleozoic era, that is, more than 250 million years ago, and for part of the following Mesozoic era, the present-day continents were grouped in close proximity, forming a gigantic original continent ("Pangaea," a "whole earth"). Only in the Mesozoic did the continents begin to drift gradually apart.

It had long been recognized that the continents of the Southern Hemisphere, including the Indian subcontinent, exhibited many common characteristics of geological development, while differing markedly from the northern continents. Twenty years earlier, Eduard Suess had proposed the name "Gondwana-Land" for this supercontinent. ("Gondwana" would actually be a more correct term since it means "land of the Gonds," which is the name of a region in India.)

Suess did not realize that anomalous early climatic conditions could also be explained by the existence of Gondwana and that Wegener's theory of continental displacement would offer a tempting solution to a climatic puzzle.

If the continents are not fixed, but rather have been displaced over the course of time, then a whole series of questions raised by the distribution of plants and animals both today and in early geologic time can be easily answered. The striking kinship of the South American and African fossils mentioned above and the close floral and faunal relationships of other areas, which are widely separated today by oceans, are problems which have long interested plant and animal geographers as well as paleontologists.

One attempt at an explanation, which was cited frequently both before and after Wegener, was the hypothetical construction of connections in the form of "land bridges," for example between "Africa-Europe" and America. The classical paleogeographical reconstructions of ancient land forms contained such land bridges, which were given names like "Archhelenis," and which were assumed to have sunk into the ocean at some later time. But the map drawn by Antonio Snider-Pellegrini in 1858 (see page 78) already offered a completely different solution, simply by pushing Africa-Europe and America together; and that is the theory which Wegener set forth in his much more detailed and well-founded exposition. It is, therefore, not surprising that a number of animal and plant geographers were especially happy to support the acceptance of continental drift and that in 1977 the Viennese paleontologist Erich Thenius dedicated his book *The Oceans and Continents in the Course of the Ages* (*Meere und Länder in Wechsel der Zeiten*) to the memory of Alfred Wegener.

Thus, the theory of continental drift has had a profound influence on our conceptions of early climatic conditions and the cause of the great climatic fluctuations, for the drifting of the continents also often entailed their moving into a different climatic zone. We have already touched on that point in connection with the Gondwana problem, and it will occupy our attention in greater detail in the rest of this chapter. In later

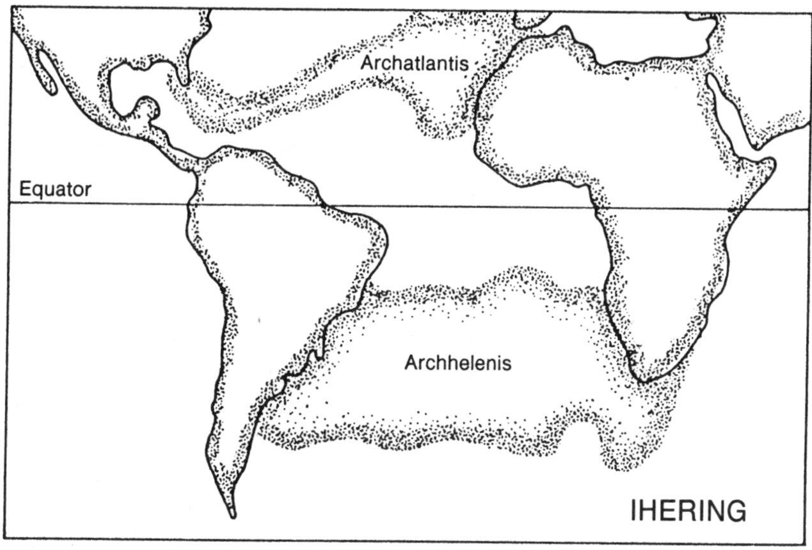

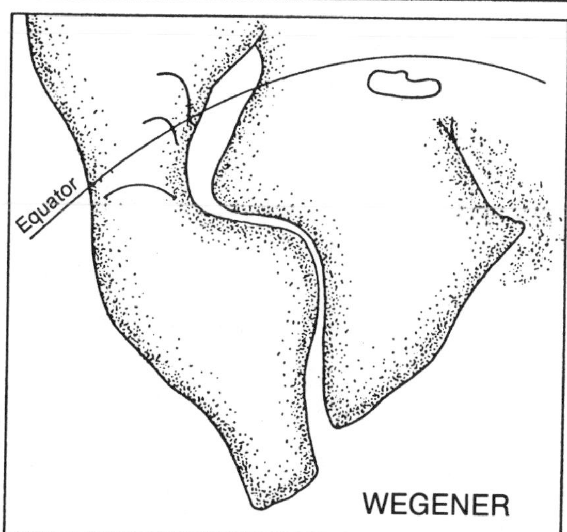

"Two world views": Ihering's and Wegener's. Explanation of floral and faunal similarities among the continents by means of either land bridges which later sank (Ihering) or continental drift (Wegener; the drawing corresponds to his reconstruction of the Cretaceous period, 1924).

chapters we will analyze what effect the drift hypothesis produced on the earth sciences after Wegener presented it in 1912.

The ideal father-in-law, Wladimir Köppen

> *Everyone who met with Wladimir Köppen always left with the feeling of having met one of the most impressive personalities in our circle of meteorologists. I had the good fortune to be in Köppen's home at an important moment in September 1908, in connection with the meeting of the German Meteorological Society in Hamburg. It happened that Wegener was there, too; thus, I encountered two men who were consumed by their scientific drive and who were spiritedly debating problems of meteorology and climatology. What was important to Wegener was the chance to present his views to this most discerning of critics, but Köppen, the mature, respected scientist, was also clearly intent on learning and absorbing what his younger colleague had to say.*[15]

One of the strokes of good fortune in Wegener's life was to have had Wladimir Köppen as his father-in-law. Köppen was already mentioned briefly above as the father of Wegener's wife, Else, and as a member of their household in Graz. But he was much more than a devoted family member: he and his son-in-law enjoyed an ideal kind of scientific collaboration. Both meteorologists, they complemented each other exceptionally well, in spite of, or, perhaps, because of their differences. Else Köppen Wegener wrote a biography of her father as well as of her husband. "Naturally, it is much easier to write about one's father than about one's husband," she responded in 1977 to the question as to which book had been easier to write. Her writing of Köppen's biography was also greatly facilitated by the detailed account Köppen himself kept of his life through 1903.

The well-known climatologist, Wladimir Köppen, Wegener's father-in-law and collaborator, at the age of seventy-eight.

Wladimir Köppen came from a German family which had lived in Russia for three generations, his Russian first name reflecting the family's loyalty to the Russian state they served. In 1786, during the reign of Catherine II, his grandfather, a doctor, had been appointed an official government physician in Kharkov. Wladimir Köppen was born in St. Petersburg in 1846 and attended Russian schools there and also in the Crimea. In 1866 he began his study of the natural sciences in Heidelberg, where he published several brief meteorological studies and where he received his doctorate in 1870 with a dissertation (published in Moscow) on the effects of heat on the growth of plants.

After working for a short time at the Central Observatory in St. Petersburg in 1875, he came to the German Marine Observatory in Hamburg through the efforts of its director, Georg Neumayer, who had been responsible for its recent rebuilding. Here Köppen first became the head of the weather service and then the "meteorologist of the observatory" (that is, "a department head without a department"), a position which offered the opportunity to do the kind of wide-ranging scientific research he desired. He stayed at the observatory for forty-four years, until 1919, when Alfred Wegener became his successor; and during that period, as well as during his long "retirement," he wrote a large number of scientific papers. The list of publications included by Else Wegener in her biography of her father numbers no less than 526 items! Of course, most of these are short notes, "thanks to which Köppen was able to add variety and currency to the learned journals." Most of them appeared in the *Meteorologische Zeitschrift* and the *Annalen der Hydrographie und Maritimen Meteorologie*, the observatory's own publication. But among Köppen's writings there are also studies of fundamental importance: a new classification of the earth's climates, which in a later, expanded form still occupies a "place of preeminence," and, most importantly, the book he wrote with Wegener, *The Climates of the Geological Past* (*Die Klimate der Geologischen Vorzeit, 1924*).

Continental drift and the climate of the past

Our meager knowledge of the climate of the past has been recently given the imposing name "paleoclima-

> *tology." But if we read the books devoted to this new science, we are faced with a confused collection of facts, whose organizing principles are still lacking.*
>
> —WLADIMIR KÖPPEN[16]

We must discuss Köppen's and Wegener's book on the climates of the past in some detail, since it bears directly on Wegener's theory of continental drift. The first chapter deals with "climatic witnesses": the rocks and the fossil plants and animals from which one can draw some conclusions about ancient climates. It is thus a review of already recognized facts. But in the second chapter, a completely new version of climatic history begins. The fossil evidence of the various epochs is not simply recounted but rather organized on world maps which Wegener constructed on the basis of his drift hypothesis. The result is a marvellously simple picture of the climate of the past, the "paleoclimate," because the drifting of the continents, of course, often entailed their displacement into a new climatic zone. The fossils which bear witness to climatic conditions, such as the fossil moraines in India or petrified giant trees in Spitzbergen, now suddenly found the geographical location that corresponded to "their" climate. Thus, "bewildering confusion" was resolved into a simple picture. Indeed, one could now attempt to deduce the positions of the equator and the magnetic poles in the past. Obviously, the movement of a continent produced a change in the position of the continent relative to both poles and to its longitudinal and latitudinal network. Thus the poles wander as well, even if only apparently: "relative" polar wandering. Whether or not substantial changes in the position of the earth's axis actually occur is another question. Wegener believed that a wandering axis was possible and considered it to be a cause of the drifting of the continents. Later on, Köppen and Milankovitch delved more deeply into that possibility. Early attempts in the seventeenth

Die Klimate
der geologischen Vorzeit

von

W. Köppen und A. Wegener
Meteorologe der Seewarte a. D. o. Prof. a. d. Universität Graz

Mit 1 Tafel und 41 Abbildungen im Text

Berlin
Verlag von Gebrüder Borntraeger
W 35 Schöneberger Ufer 12a
1924

The title page of The Climates of the Geological Past, *a pioneering work in paleoclimatology.*

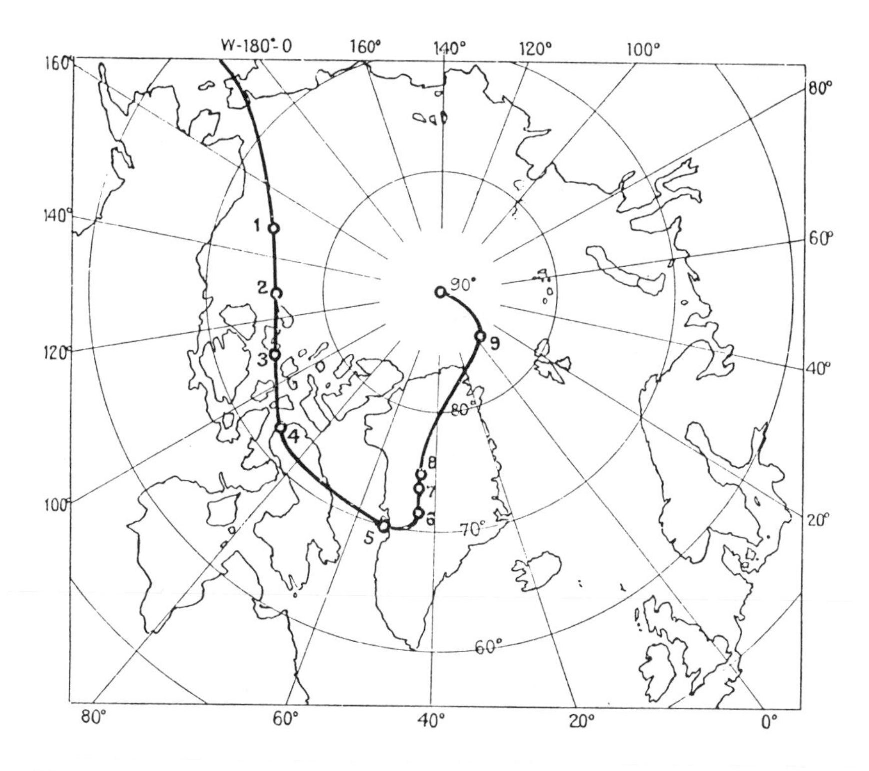

The movement of the magnetic North Pole, beginning in the Miocene period, 1, to the present. From Die Klimate der Geologischen Vorzeit.

century to explain climatic changes during the earth's past also proceeded from the assumption of an actual change in the position of the poles at rotation.

An outstanding achievement of the drift hypothesis, recognized as such even by otherwise sceptical scientists, is the explanation of the existence of Permocarboniferous glaciations. Traces of great glaciers, existing during the Carboniferous and Permian periods of 250—300 million years ago, were first discovered on the Indian subcontinent in 1856 and then in all of the southern continents: glacial erosion, moraines, and so forth, along with flora typical of a cool climate, such as *Glossopteris*, a genus of primitive ferns. The distribution of these features over such a wide area and the evidence they offered of the previous existence of an ice sheet can be reconciled neither with present-day climatic conditions nor with the positions of the southern continents relative to one another. Wegener resolved the difficulties by "pushing together" all of these continents, as well as the northern ones, to form a new supercontinent, which he named "Pangaea." He also moved the South Pole to "South Africa-Australia." Now the traces of glaciation are all united in the region of the South Pole, and the Carboniferous forests of that time in Europe and North America, according to Köppen and Wegener, indicate the location of the equatorial zone. To be sure, the Carboniferous flora do not constitute direct proof of the location of the equator, as the two men assumed. They based their conclusions on the paleobotanical research of Henry Potonié, a nineteenth-century geologist.[17] But the possibility of this configuration can easily be imagined.

In Köppen's and Wegener's reconstruction of this period, Antarctica is also included in the "South Polar region", as an empty space, without any notation as to glacial climatic evidence, of which nothing was known at the time. Only in 1960 was the first evidence discovered which validated their hypothesis. Thus, the paleoclimatic classification which Köppen and Wegener had postulated in 1924 was confirmed thirty-six years later.

No one before Köppen and Wegener had offered as thorough and consistent an interpretation of the earth's climatic history. Wegener's bold flights of imagination and Köppen's broad knowledge combined to build a convincing conceptual structure. But it is also true that this structure was not entirely

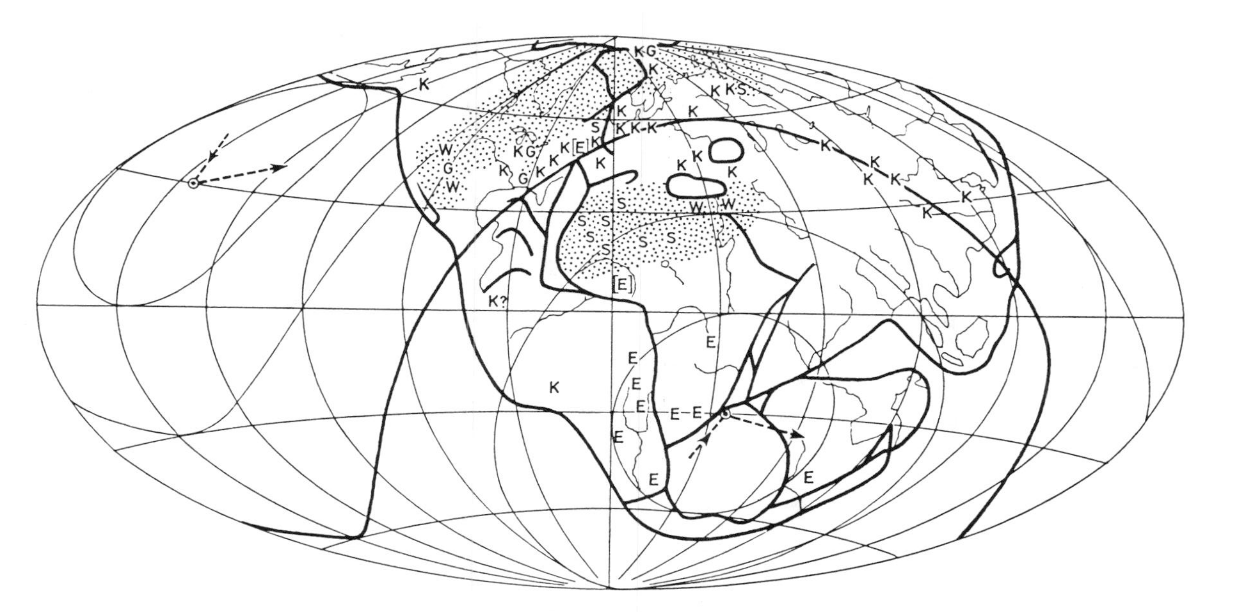

The earth during the Carboniferous period: Pangaea, as first published in Köppen-Wegener, 1924. For the southern continents, glacial deposits (E) are found. G (gypsum), S (salt), and W (desert sandstone) characterize the dry regions, and K (carbon) mainly the equatorial zone. The arrows show the movement of the North and South Poles; the latitude lines at 60 degrees and 30 degrees, and the equator in the Carboniferous are also drawn. Some of the details no longer fit with our current knowledge, but in principle the reconstruction is remarkably good.

sound, and that some questions could be answered without recourse to continental drift, or by postulating the slow movement of the earth's crust as a totality rather than the drifting of individual continents. In a stimulating book, *The Equator Question in Geology (Die Äquatorfrage in der Geologie,* 1902), Damian Kreichgauer, a physicist, advanced the latter argument. But this sort of drifting, which does not allow the continents to change position relative to each other, explains very little.

The radiation curves of Milankovitch

In this way the Ice Age received its calendar. Its course obeys the strict laws of astronomy and the stylus of the muse Urania, the sublime daughter of Zeus and Mnemosyne. At Köppen's invitation, I had the high honor of taking this divine muse on my arm and accompanying her into the realm of geology.

—MILUTIN MILANKOVITCH[18]

The book by Köppen and Wegener on paleoclimates contains yet another completely new solution to a puzzling geological finding. It had been known since the previous century that the major Ice Age (or, more precisely, the Quatenary Ice Age) consisted of several cold periods—the actual periods of glaciation—which were interrupted by warmer periods—the interglacial periods.[19] As far as the Alps were concerned, the scientific division of the Ice Age into several distinct periods was put on a firm footing by Albrecht Penck[20] and Eduard Brückner,[21] who were able to distinguish four separate periods of glaciation. In their magnum opus *The Alps in the Ice Age (Die Alpen in Eiszeitalter)* they noted that their research was stimulated by a prize offered in 1887 by the Breslau section of the German-Austrian Alpine Association.

Then Köppen realized that one could compare the geological evidence of the repeated alternations of warm and cold periods with astronomical calculations about the periodically changing intensity of solar radiation.[22] As early as 100 years previously, the Englishman James Croll[23] had first set down ideas of this type. But only during World War I and the following years did the Yugloslavian astronomer Milutin Milankovitch calculate precise dates and plot the data.[24] Thereby, he obtained a large number of maxima and minima of radiation, all defined as to the exact year.

The most important impetus now came from Köppen, who, together with Wegener, applied the Ice Age climatic curve that Penck and Brückner had derived from their geological observations to the astronomical radiation curves. "The first application of the radiation curves to problems of paleoclimatology was made by W. Köppen and A. Wegener," Milankovitch wrote in 1938.[25] It was also Köppen who encouraged Milankovitch to calculate the curves as far back as 650,000 years ago. In this way, the several periods of glaciation known for the Alps could be dated by year with astronomic precision. According to those calculations, for example, the last glaciation period had three maxima—at 26,000; 70,000; and 115,000 years ago. This seemed to represent dramatic progress, for, essentially, two birds had been killed with one stone. First of all, there was now an explanation for the climatic variations of the Ice Age, and at the same time a completely new method of precisely dating geological processes had become available. Wolfgang Soergel, in Freiburg im Breisgau, and his student Friedrich Zeuner[26], who later worked in London, were particularly instrumental in refining the application of Milankovitch's radiation curves.

Milankovitch described all of this, with particular emphasis on his close collaboration with Köppen and Wegener, in a witty book which he wrote for a wide audience in the form of

Milutin Milankovitch (1879–1958), the Yugoslavian climatologist. The "radiation curves" he calculated were first used by Köppen and Wegener for their chronological classification of ice age processes.

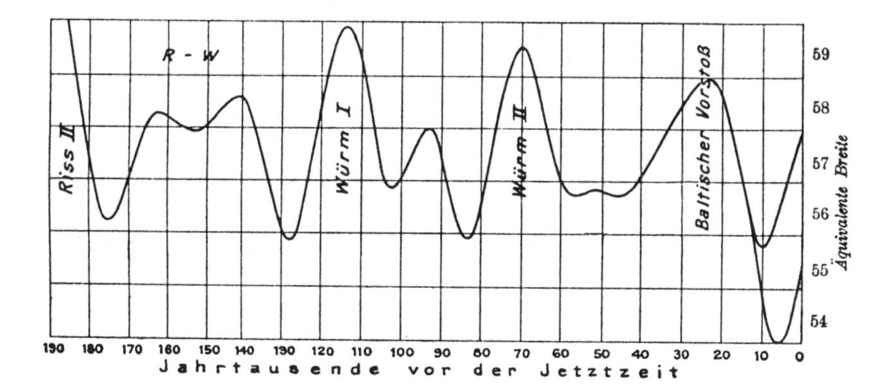

Radiation curve for the last 190,000 years, according to Milankovitch: From Die Klimate der Geologischen Vorzeit *by Köppen and Wegener.*

charming letters to an imaginary girl friend: *Through Distant Worlds and Times: Letters from an Ambler through the Universe (Durch ferne Welten und Zeiten)*, which was published in Leipzig in 1936.

Milankovitch spoke about his life's work for the last time at the International Quatenary Conference in Rome in September 1953. It was probably the first time that he had presented his findings to a large international audience. Unfortunately, his presentation remained incomplete, since the vigorous moderator of the session, Richard F. Flint,[27] who was used to the strict customs of American conferences, cut off the seventy-four-year-old speaker after his allotted time of twenty minutes had elapsed. A few weeks later Milankovitch referred to that episode in a letter to Walter Wundt, his colleague in Freiburg and an advocate of his theory: "It didn't bother me that my talk was so inconsiderately interrupted. The applause I received from everyone in the audience convinced me that our theory cannot be denied out of existence." At another time he wrote: "The fluctuations in the radiation received by the earth over long periods of time are only *one* component of the climate of the past, but they are the most important one, and, moreover, one which is amenable to precise investigation. That is enough for me. So, I can sleep peacefully, and also die at peace, when the time comes." He died in 1958.

After 1924, the "radiation curves" played an important role in the thinking of earth scientists. Not only Quatenary geologists, but also historians of the ancient past appreciated such an exact time table. New advocates and new opponents also appeared in increasing numbers from the ranks of the "physical geographers," among whose leading thinkers was Albrecht Penck.[20] Penck was never able to feel much enthusiasm either for continental drift or for the radiation curves. Nevertheless, he greatly respected Wegener as a creative scientist and endeavored, in vain, as it turned out, to make him the successor to Alfred Merz as Director of the Museum of Oceanography in Berlin.

In conclusion, we can say today, more than a half-century after the publication by Köppen and Wegener of their book, *The Climates of the Geological Past*, that the optimistic expectations which they, along with Milankovitch and many other experts in the field, had placed on the radiation curves, were not fulfilled. The astronomical timetable was indeed correct, but its application to a geological division of the Ice Ages was mistaken. That has become increasingly evident as other, newer methods of precisely dating geological processes have become available. Obviously, the rather small fluctuations in the sun's radiation are not sufficient in and of themselves to have such large-scale effects as the Ice Ages. However, small fluctuations in climate do appear to be connected to the radiation curves, and they are thus still of some importance to Ice Age research, especially since new calculations have improved their accuracy. "At best, the Milankovitch mechanism may be responsible for the minor ripples superimposed on the major trends," as

D. M. Parkin stated in 1976. They may be *one* of the causes active in a "multilateral" creation of the Ice Ages, perhaps even as a kind of "pacemaker."[28]

Unfortunately, the radiation curves are of no use for predicting future ice ages, even if Köppen wrote optimistically in 1931, "According to astronomic data, a return of the Ice Age is out of the question for the northern hemisphere for the next 20,000 years and even for many years beyond that time."[29] In this case, his prediction must be judged unfounded and premature. (Later on, "ice age pessimism" and warnings of an impending ice age became fashionable, but that was just as premature.) For the next ice age, completely different causes and contributing factors may be involved.

Periods of glaciation and the Ice Age: Why?

> *But that is exactly what I like about this science of geology. It is infinite, ambiguous, like all poetry; like all poetry it has secrets, is permeated by them, lives within them, without being destroyed by them. It does not lift the veil, but only moves it, and through tiny holes in the fabric a few rays escape, which dazzle the eye.*[30]

In their book, *The Climates of the Geological Past*, Wegener and Köppen postulated two completely different causes for the great fluctuations of climate over the course of the earth's history and applied these causes to periods of glaciation of "medium" and "long" duration.[31] They used the radiation curves to explain the periodical climatic changes in the Quatenary era, that is, for the occurrence of several periods of glaciation within an Ice Age. Its beginnings, we can assume today, probably go back several tens of millions of years. Both Wegener and Köppen realized, however, that the radiation curves could not explain *everything*: they can be calculated back further and further (although with ever-decreasing precision), but geologists have no evidence of periods of glaciation for the "radiation minima" of the preceding 250 million years. Obviously, we must also assume that there was a special state of "readiness" for the occurrence of periods of glaciation, in order to explain the great ice ages in earth's history. Köppen and Wegener found the "primary" cause for the creation of ice age conditions in continental drift, that is, in the location of a continent in a polar region. According to modern paleomagnetic data, however, this "coincidence" (as Fairbridge[32] termed it) does not seem to suffice: presumably Antarctica was already at the South Pole long before the most recent ice age.

Thus, in spite of their optimism, Köppen and Wegener were not able to answer the question of the ultimate cause of the great glaciations. Nevertheless, today we can seriously consider the possibility that the radiation curves play only a small role and that continental drift is probably of predominant importance in the earth's climatic history.

Wladimir Köppen in retrospect

> *What a good thing it would be if every scientific man was to die when sixty years old, as afterwards he would be sure to oppose all new doctrines.*
>
> —CHARLES DARWIN, AUTOBIOGRAPHY[33]

This facetious comment by Darwin clearly does not apply to Wladimir Köppen. Although at first, Köppen rejected his son-in-law's ideas on continental drift, he soon became one of Wegener's most enthusiastic supporters. "My father always carried a small globe in his coat pocket, in order to be able to think about this idea at anytime," Else Wegener reported. When Köppen and Wegener's book appeared, Köppen was seventy-eight years old, testimony to the amazing vitality of his intellectual powers into old age, and to his capacity for assimilating completely new and revolutionary views. Ten years after Wegener's death, the ninety-three-year-old Köppen compiled a series of "Supplements and Corrections" to their book. In her biography of her father, Else Wegener wrote:

> In mid-June (1940), much against my will, I had to send a telegram for him to the publishing firm in Halle which was in the process of printing his "Supplements": "Please send proofs at once, am dying." But when the proofs arrived, I had to read them. . . .In the early morning hours of June 22, 1940, he died peacefully in his sleep.

6

"WANDERING CONTINENTS": THE FIRST FORTY YEARS

The way in which a new scientific truth usually becomes accepted is not that its opponents are persuaded and declare themselves enlightened, but rather, that its opponents gradually die off and the following generation grows up accepting the truth from the start.

—MAX PLANCK[1]

“WANDERING CONTINENTS”: THE FIRST FORTY YEARS

Wegener first published the hypothesis that made him famous in 1912, when he was only thirty-one years old. It marked his entry into a field that was completely new to him, and he used the term “geophysics”[2] to describe this field. Indeed, *geophysical* considerations about the “floating” continents and the forces that set them in motion did play a major role initially. But from the beginning, the evidence obtained from *geological* observations was of primary importance; after all, it was the geologists who dealt with the question of the origin of mountain ranges, which, in the final analysis, also belonged to Wegener’s hypothesis.

Wegener’s first paper and first book on continental drift

In the following years geologists tended to pay much more attention to Wegener—as far as this hypothesis was concerned—than did geophysicists. In general, Wegener’s relationship to geology was a cool one, as was made evident on more than one occasion. And yet his first paper, entitled “The Geophysical Basis of the Evolution of the Large-scale Features of the Earth’s Crust (Continents and Oceans),” was presented before a geological forum, at the annual meeting of the *Geologischen Vereinigung* (Geological Association) in Frankfurt am Main on

January 6, 1912. That day was a historic one for the earth sciences, no matter what one thinks of continental drift.

Although up to that time Wegener had written only on climatology, it is easy to explain why he appeared before a *geological* association. The chairman of the Vereinigung at that time was Emanuel Kayser, who was a colleague of Wegener's in Marburg. The association had been founded only two years before as a sort of complementary body to the venerable, tradition-bound Deutschen Geologischen Gesellschaft: not as a rival, it was emphasized, but rather to foster the study of "general geology", which was given short shrift in the more regionally oriented journal of the Deutschen Geologischen Gesellschaft. In fact, however, the new Geologischen Vereinigung also functioned as an invigorating counterpart to its older sibling organization, infusing new energy into the profession under the imaginative leadership of Gustav Steinmann.[3] Later, the impulsive, high-spirited Hans Cloos put his stamp on both the new association and its journal, *Geologische Rundschau*, which gained international stature under his leadership. Eventually, the Geologischen Vereinigung became the largest professional society of German geologists.

Obviously, this meeting was the appropriate place for Wegener to present his new hypothesis on continental drift. Of course, that does not mean it met with general acceptance. Although there was evidently no opposition expressed at the time (the brief report on the meeting states that "there was no discussion due to the advanced hour"), the reaction over the next several years—and even over the next several decades—was overwhelmingly negative.

The paper was published in the same year in the *Geologische Rundschau* and also in somewhat expanded form in *Petermanns Geographischen Mitteilungen*. It is worth noting that the paper in the *Geologische Rundschau* is the only one that Wegener ever published in a German *geological* periodical. On the other hand, he often published articles in *geographical* journals.

Wegener's participation in the Greenland expedition of 1912–1913 and the outbreak of World War I delayed the publication of a book-length account of his hypothesis until 1915, when *The Origin of Continents and Oceans (Die Entstehung der Kontinente und Ozeane)*, his major work, first appeared in print. A second, revised edition followed in 1920, a third edition in 1922, and finally a fourth edition in 1929, by which time the book had grown to 231 pages from the original ninety-four pages. The fourth edition was reprinted as the fifth edition in 1936, as the sixth edition in 1941, and, for the last time, in 1962. The third edition was also translated: into French, English, and Spanish in 1924, into Russian in 1925, into Swedish in 1926, and into Italian in 1942; the fourth edition was also translated into English in 1966. Surely, there are few original works in the natural sciences which have achieved such a wide dissemination or which have provoked so much discussion, despite widespread skepticism and opposition.

From the very beginning, Wegener was remarkably confident in the truth of his hypothesis. On December 31, 1911, he wrote to his doubting father-in-law (see figure, p. 183),

> If it turns out that sense and meaning are now becoming evident in the whole history of the earth's development, why should we hesitate to toss the old views overboard? Why should this idea be held back for ten or even thirty years? . . .I don't think the old ideas will survive another ten years.

Of course, this prediction was not borne out.

The reason for the exceptional attention Wegener's ideas attracted is, at least in part, that their basic premise can be readily understood even by the lay person. For example, Hans

Cloos summarized the hypothesis with this lyrical but simple description:

> It placed an easily comprehensible, tremendously exciting structure of ideas upon a solid, scientific foundation. It released the continents from the earth's core and transformed them into icebergs of gneiss on a sea of basalt. It let them float and drift, break apart and converge. Where they broke away, cracks, rifts, trenches remain; where they collided, ranges of folded mountains appear.[4]

Much of that explanation can be easily understood by a bright twelve-year-old child, for example.

A prophet without honor in his own country

On the other hand, the argument advanced by this outsider to geology was so revolutionary that most geologists rejected it out of hand. Many of them refused to take it seriously and simply ignored it. Others, while not accepting the hypothesis, were fascinated by Wegener's bold flight of imagination. Hans Cloos was one of this group, as we already discussed in the chapter on Wegener's experience in Marburg.

Certainly, there were many points in Wegener's argument that could be easily refuted, whether by geophysicists, geologists, or paleontologists. In many respects continental drift seemed superfluous (for example, if land bridges had existed), especially when even Wegener was unable to explain satisfactorily the mechanism that set the continents in motion. With this in mind Hermann von Ihering[5] gave the last chapter of his 1927 book on the history of the Atlantic Ocean the provocative title, "Two World Views: v. Ihering and Taylor-Wegener." For Ihering, Wegener's idea was "the fantasy of a geophysicist. . .that would pop like a soap bubble."

There were others who were unsparing with polemical remarks and referred, for example to the "delirious ravings of people with bad cases of moving crust disease and wandering pole plague" (Fritz Kerner-Marilaun,[6] a respected Austrian paleoclimatologist, in 1918). Other amusing examples can be found in a paper by Max Semper,[7] of Aachen, who had previously published thoroughly solid research on the climate of the Tertiary period. The proof of "the reality of continental drift," he writes, "is undertaken with inadequate means and fails totally." He advised Wegener "not to honor geology with his presence any longer, but to look for other fields that have so far neglected to write above their door, 'Holy Saint Florian, spare this house—set other ones on fire'." That is obviously the attitude of a scientist who is unable to see that even an outsider, a "non-geologist" in this case, can cultivate a new field in geology successfully, or, in some circumstances even more successfully than a geologist.

In spite of this opposition, there was continued interest in continental drift, as the repeated publication of the book in new editions attests. In them, Wegener continued to add new arguments, to drop others, and attempted to refute the counter-arguments of his critics.

It was actually in Germany that Wegener found the least acceptance—"a prophet without honor in his own country." It is true that his hypothesis did have a few prominent champions, for example the Heidelberg geologist Wilhelm Salomon-Calvi,[8] who even suggested a special term for the "horizontal movement of the continents": *epeirophoresis*,[9] which, however, never became widely accepted. But other geologists who were even more respected and influential maintained their opposition to Wegener's hypothesis. Most important among them were the two towering figures of German geology in those years, Hans Stille[10] and Hans Cloos.

In Stille's numerous papers on tectonics—he published detailed studies on the Atlantic and Pacific basins, focal points of the continental drift hypothesis—Wegener's name is scarcely mentioned. "He was always opposed to those 'mobilist' ideas. Even very early in his career he was more inclined toward a 'fixist' position, with reference to the eternal nature of the continents and oceans"—said one of Stille's former students, Andreas Pilger, who spoke at the commemoration of Stille's one-hundredth birthday in 1977, almost ten years after his death.[11] Very honestly, the speaker added what could not be said during Stille's lifetime: "Given what we know today, it must be said that Wegener's mobilism, rather than Stille's fixism, pointed us in the right direction." However, Stille's unbending opposition turned out to have a long-lasting influence, since there existed a large "school" of his former students who revered him as the "master" (which in other respects he truly was), often unconditionally. It had been much the same 150 years earlier with Abraham Gottlob Werner and his Freiburg school of Neptunism. "Geologists, by nature perhaps too conservative, often are too willing to follow established leaders," Robert Scholten wrote.[12] But the other natural sciences could offer similar examples of blind allegiance to authority.

The negative view that Cloos also took toward Wegener's hypothesis had far-reaching effects, mainly due to the great renown that Cloos, like Stille, enjoyed beyond the boundaries of his own country. As a result, "In Europe and particularly in Germany, Alfred Wegener's ideas were virtually filed away and forgotten for three decades," wrote Hans Georg Wunderlich, a far-sighted but short-lived geologist of the younger generation.[13] At the same time, Cloos was an entirely different sort of man than Stille. Cloos was not able to support Wegener's ideas, but that did not prevent him from finding them "brilliant" and treating them in detail in his introductory textbook in geology. Nothing illuminates his completely unprejudiced attitude better than a proposal he made at the first paleoclimatology meeting of the Geologischen Vereinigung in 1950 in Cologne: He proposed that the Gustav Steinmann Medal, a significant award, should be given to Else Wegener in posthumous honor of her husband. Cloos' suggestion was received without enthusiasm: Instead, the medal was awarded to Hans Stille.

The continental drift debate in New York in 1926

In North America the battle over continental drift, while somewhat different in detail, came to a similar end during the 1920s. A huge symposium was held in November 1926 in New York, at which Wegenerites and anti-Wegenerites set forth and defended their positions candidly, and debated the issues quite sharply. The meeting was entitled "Theory of Continental Drift. A Symposium on the Origin and Movement of Land-masses both Intercontinental and Intracontinental, as Proposed by Alfred Wegener." The symposium was by no means the first of its kind; there had been a similar one, for example, at the 1922 meeting of the British Association for the Advancement of Science in Hull. But this meeting in New York represented a high point in the controversy surrounding continental drift. A series of eminent earth scientists (even if most were "endowed more with influence than with vision"[14]) presented papers. It is particularly interesting that the sponsor of the symposium was the influential American Association of Petroleum Geologists, a society directed entirely toward the "applied" side of geology. But, as De Golyer stated in the prefatory note to the proceedings of the conference, "There is little in the realm of geology that is unimportant or solely of academic interest to the economic geologist of today." The actual initiator of the symposium was the learned and successful Dutch oil geologist W.A.J.M. van Waterschoot van der Gracht, a doctor of law who had also been a mining engineer and was now active in the United States. Wegener did not attend the symposium,[15] but did contribute a brief (and not very interesting) note to the published proceedings.

Van der Gracht was a strong supporter of the continental drift hypothesis. In an outstanding introductory address and in a summation of the papers contributed to the symposium he set forth objectively the arguments for and against continental drift. Even though he was favorably disposed toward it, he did not deny that many details of the theory, particularly the mechanical and physical explanation, would require much more research. That opinion was shared by the few others who supported Wegener. Most of the participants rejected the theory out of hand, at times quite caustically. Wegener lost the deciding battle in North America, too, if one may put it that way, and the result was similar to the situation in Germany. Frederic J. Vine, two generations younger than some of the participants in the New York symposium, said much the same thing in 1977—for forty years no American geologist considered seriously the possibility of continental drift.[16] Of course, that is something of an exaggeration, but is nevertheless true for most North American geologists. "Continental Drift, Ein Märchen" (A Fairy Tale), the title of an article by Bailey Willis,[17] expresses an attitude similar to Kerner-Marilaun's; and as recently as 1966, Pascual Jordan also described continental drift as the "geophysicists' favorite fairy tale."[18]

Convection currents: Ampferer, Schwinner, Holmes

Among the critical observers of the New York symposium was Arthur Holmes, a well-known professor of geology at Edinburgh,[19] who referred to the meeting as "famous, if not infamous." In his opinion, the discussions were not only "rough" but also only tangential to the heart of the controversy: it was true that many of Wegener's supporting arguments had been refuted, which was not difficult to do, but not his major idea that the continents can drift. Chester R. Longwell shared that view: "The goal of scientific endeavor is to learn the truth of nature and not to win debates," he wrote in 1944.[20]

Holmes was no blind disciple of continental drift. "I have never succeeded in freeing myself from a nagging prejudice against continental drift; in my geological bones, so to speak, I feel the hypothesis as a fantastic one."[21] Nevertheless, "I have been deliberately careful not to ignore the very formidable body of evidence that has seemed to make continental drift an inescapable inference." "Sterility," Holmes once replied to a caustic remark of Vladimir Beloussov, a Russian opponent of continental drift, "is about the least appropriate and most unjust epithet that could be applied to the continental drift concept."

One of the weakest points in Wegener's argument was his explanation of the mechanism which causes the continents to drift. Holmes believed that the forces which were supposed to be operative were "hopelessly inadequate." Early on, Holmes himself had postulated slow convection currents in the viscous magma of the mantle as the driving force and thus had indicated the path that plate tectonics would later follow. But he was not the first to describe this possibility. Two Austrian geologists who were thoroughly acquainted with the empirical evidence of a complicated tectonic structure for the Alps had lit upon that idea before Holmes. One of them was the Tyrolean Alpine geologist Otto Ampferer,[22] who had advanced his theory of "undercurrents" as early as 1906. Later on, in 1925, Ampferer took issue with Wegener on the theory of continental drift, in particular as far as the mechanism of motion was concerned. Ampferer rejected the outside forces that Wegener had postulated, stating, "It seems to me that currents in the molten interior of the earth must be contributing factors at the very least." His schematic drawing depicting this process is amazingly similar to modern concepts in plate tectonics.

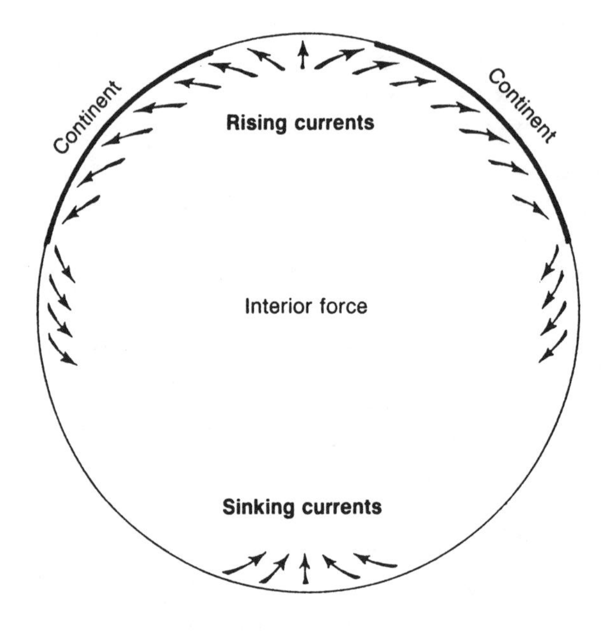

Convection currents as the impetus for continental drift—diagrammed in this drawing by Ampferer as early as 1925.

The second Austrian to introduce the idea of convection currents was Robert Schwinner[23] of Graz (see also Chapter 7, page 127). However, for many years no one took the astute observations of these two men any further. Wegener did mention Ampferer in 1920, and Schwinner in the 1929 edition of his book, but did not take advantage of the fundamental possibilities their ideas offered for his hypothesis.

That he did not mention Schwinner is actually quite surprising. After all, Schwinner was Wegener's colleague for many years at the small university in Graz, and, although he may have been difficult personally, he was a creative thinker, thoroughly at home in both geology and geophysics. Yet there seems to have been no close contact between the two men. In his *Textbook of Physical Geology (Lehrbuch der Physikalischen Geologie)*, Schwinner offered a detailed and objective presentation of Wegener's hypothesis; he did draw attention to its weak points, but in this he was no different from other critics. Perhaps Schwinner felt a bit reserved in relation to Wegener, who was his junior and yet had advanced further in his academic career; on the other hand, perhaps Wegener perceived Schwinner as a disconcerting critic. That would explain their cool relationship. At any rate, it is both remarkable and regrettable that Wegener, Ampferer, and Schwinner all contributed fundamental ideas to the theory of continental drift independently of one another, while living in the same small Alpine area of Austria at the same time.

Earlier, in Marburg, Wegener's relationship with the geologists had been very good: "I require a lot of material for the work I'm doing. The geologists hunt up everything I need and give it to me, so that I am relieved of nine-tenths of the work. Otherwise I would need months to get as far as I have now" (from a letter to Wladimir Köppen, January 29, 1912).[24] In contrast, in Hamburg and later in Graz, Wegener made few attempts to interact, either with Schwinner, or with another geological colleague, Franz Heritsch. To that extent one would almost have to agree with the statement in Schwinner's textbook that the continental drift hypothesis had originated "far from geology."

However, the idea of "drifting continents" may have occurred to Wegener for this very reason—that he had *not* studied geology. In a similar context, Anthony Hallam, in his book *A Revolution in the Earth Sciences*, discusses the work of John Dalton, also an outsider, and, coincidentally also a "meteorologist", who revolutionized the conceptual world of chemistry with his theory of atoms.

Alexander Du Toit (1878–1948), important geologist and supporter of Wegener in the early years of the drift hypothesis.

Africa forms the key[25]

The most important proponent of continental drift during those early decades of the twentieth century was Alexander L. Du Toit, a descendent of a Huguenot family that had emigrated via Holland to the Dutch Cape colony in the seventeenth century. Du Toit, the "most-honored geologist in South Africa" with five honorary doctorates from that country alone, was one of Wegener's staunchest and most respected supporters. His book, *Our Wandering Continents: An Hypothesis of Continental Drifting*, published in 1937, is a standard work, and even the first edition contained the significant dedication: "To the memory of Alfred Wegener." The book shows Du Toit as an "iconoclast," as S.H. Haughton put it in his obituary note,[26] who warned geologists "against barricading themselves behind conservatism and the same old ideas," and who demanded "a reexamination of observations and 'congruences' in the light of new theories."

It is hardly coincidental that it was an African who entered the fray so energetically in support of Wegener's hypothesis and used the words, "Africa forms the key," as an epigraph to his book. Africa does, in fact, occupy a key position in several respects. The evidence of Permocarboniferous glaciations, mentioned earlier, is seen there with marvelous clarity. This is also true of the other southern continents, as Du Toit knew from his travels. Anyone who goes there is immediately struck by the unexpected sight of what was once icy wastelands in those now torrid regions, and, of course, is also struck by the thought that a glaciated Gondwana much larger than today's Antarctica is most easily accounted for by Wegener's reconstruction.

Geodetic evidence: Wegener's error

Wegener had put particular emphasis on *one* possible proof of continental wandering and had even regarded it as decisive: the demonstration of movement by means of precise measurements of geographic longitude. A substantial shift seemed likely for Greenland, in particular, and thus, perhaps easily subject to direct verification. In 1823 Edward Sabine[27] had measured the longitude of a small island near the east coast of Greenland, using the crude astronomical methods of his time. In a letter Wegener wrote in 1912 he suggested that it would be useful to obtain new measurements of Greenland's longitude, in order to determine whether the island had moved in the intervening ninety years. In the last edition of his book on continental drift he devoted twelve pages to "geodetic arguments" alone, and he went into them again in an appendix. Of course Wegener, who had become thoroughly familiar at astronomic-geographic position finding in the course of his studies at the university, knew how uncertain longitudinal measurements are, which at that time were based on lunar observations. Better methods to measure the longitudinal differential between Europe and North America, for example, became available only when the chronometer was replaced by radiotelegraphy.

Although changes in the distance between two continents, if they occur at all, obviously occur only extremely slowly, Wegener nevertheless estimated the increase in distance between Scotland and Greenland at 18 to 36 meters per year, based on his drift hypothesis. Assuming that measurements separated by a number of years were used, that might allow a verdict either for or against his drift hypothesis, even if the measurements had been taken with the earlier, imperfect methods. Wegener believed that the results available to him allowed one to see at least a tendency toward the westward drift of North America; and in 1929 he even saw a confirmation of his theory in the latest measurements. Unfortunately, the longitudinal determinations he used for comparison were so error-ridden that they cannot be used as any sort of proof.

In fact, modern estimates for the rates of continental movement are on the order of a few centimeters per year—only a fraction of those that Wegener had postulated. Even with modern, highly refined methods of measurement using satellites, it will be many years before any usable results for changes in the distance between two continents become available. Here again Wegener was much too optimistic.

7

THE REBIRTH OF CONTINENTAL DRIFT: PLATE TECTONICS

Nevertheless, it does move.

—GALILEO[1]

The Rebirth of Continental Drift: Plate Tectonics

At the time of Wegener's death in 1930, and for decades afterwards, continental drift was scarcely a topic of scientific discussion. The prevailing opinion was that the hypothesis could not compete with the traditional views of the structure and development of the earth's crust. A number of scientists felt that the idea was simply not worth discussing.

However, Wegener's theory was not dead, by any means. A few prominent scholars such as Du Toit continued to fight for it tirelessly. Further support came from research in tectonics, which seemed to point in the same direction. While Wegener had drawn heavily on that research in the last revision of his book, its value was not generally recognized until much later. That was true, for example, of a basic work called *The Tectonics of Asia* (*La Tectonique de l'Asie*) by the Swiss Alpine geologist Emile Argand.[2] The book was finally translated into English in 1977, fifty-five years after its initial publication, and thus became accessible to the Americans, who hailed it as a "prophetic masterpiece." Another guidepost was the theory that currents in the magma are the driving force for continental drift, a concept which had been developed by two Austrians, Otto Ampferer and Robert Schwinner. However, their ideas were largely ignored, even in German-speaking countries; Wegener himself was slow to recognize their importance.

Oceanographic research and paleomagnetism

A complete about-face in scientific opinion began after World War II, when continental drift experienced a rebirth and, in the form of plate tectonics, became the basis of modern geology in the 1960's. A new foundation for the theory was laid mainly by modern oceanographic research and new paleomagnetic methods. Continental drift was "resurrected by the same discipline that had condemned it earlier," wrote Robert Scholten in 1978.[3]

The geological and geophysical exploration of the oceans, especially the deep sea, developed rapidly after the war, primarily in the United States and the Soviet Union, as well as in several smaller nations. With improved geological, paleontologic, sedimentary-petrographic, geophysical, and radiometric methods, an enormous volume of new data about the ocean floor and its underlying layers became available. A high point was the research carried out by the American vessel "Glomar Challenger," which was able to take drillings over 1000 meters down below the floor of the ocean, and thus allowed scientists to reach unexpected but well-founded conclusions about the structure of the oceanic crust. In Wegener's day, almost nothing had been known about the layers underlying the ocean floor.

A second powerful stimulus was the development of a paleomagnetic method which enabled scientists to reconstruct, for any given location, the former magnetic field of the earth. The method is based on the fact that lava "locks in" the orientation of the magnetic pole at the time it hardens, and that orientation can be measured today. In addition, the date at which the lava hardened can also be determined. Over the past generation, a great number of studies of rocks from all continents and of all ages have been completed. The measurements have led to the following conclusion: at one time the earth's magnetic poles had an orientation at many points which was very different from their present ones. And if we assume that the magnetic poles were located near the geographical poles, as is the case today, the earlier orientation of the geographic poles and their geographic latitiudes can be deduced. Given these assumptions—which are still controversial in some of their details—we have direct proof of continental drift. "Others hail the magnetic records as the long-sought Rosetta stone, a key to a major geological puzzle."[4]

Plate tectonics and Wegener's theory of continental drift

Within a few years, the wealth of new data had led to a new synthesis, *plate tectonics*. The development of this theory, which took place mainly in America, was shared equally by geologists and geophysicists, who collaborated so closely and successfully that it is frequently impossible to point to individual contributions. Among the often-cited founders of plate tectonics are, to name only a few: Robert S. Dietz, Harry H. Hess, Drummond H. Matthews, Frederic J. Vine, and J. Tuzo Wilson (see the bibliography). As mentioned above, a few assumptions basic to the theory had been published by others years earlier. But there are many other instances of unjustly forgotten pioneers on the verge of great discoveries. Sometimes they appear on the scene too early.

At the center of this "new global tectonics," as it is called, is once again the postulate that the continents are moving, but now that assumption is based on much better observations and measurements than had been possible in Wegener's time. The paleomagnetic studies offer particularly strong support. While their results are not always unequivocal, on the whole they offer weighty evidence of continental drift—evidence which, like much other information, was not available to Wegener.

Thus, plate tectonics is also a theory of continental drift. Many of Wegener's paleogeographic reconstructions have been confirmed, some quite unexpectedly; for example, his interpretation of the origin of the Afar triangle in Eritrea. Yet even though plate tectonics developed out of Wegener's hypothesis and often reaches the same conclusions, one cannot equate the two theories. In their details there are fundamental differences: In plate tectonics, it is not only the continents which are moving, but rather much larger units of the earth's crust ("plates"), which include parts of the oceans as well as the land masses. The plates—six larger and several smaller ones—move on deep-lying strata of the upper mantle, the driving force being slow currents in the viscous magma of the mantle, the "asthenosphere,"[5] as Ampferer, Schwinner, and especially, Holmes had suspected. In the process, the magma wells up within narrow zones and builds up the mid-ocean ridges, such as the mid-Atlantic Ridge, of which Iceland is a part (see page 126).

The Atlantic Ocean remains the prime example of all these processes. According to the theory of plate tectonics, North and South America and "Africa-Europe" were once joined but began drifting apart during the Mesozoic. Up to this point everything agrees with Wegener's hypothesis, although his chronology of the drift process was incorrect.[6] But for Wegener, the mid-Atlantic Ridge was not particularly important; it merely marked the site of the separation of the continents. He saw it as dead, an ancient ruin, a "waste product" (that view was already discussed in Chapter 4 with respect to Iceland). But for plate tectonics, the mid-Atlantic Ridge is very much alive: it is the suture between two plates at which new ocean floor is constantly being created by the influx of magma from the depths at the same rate as the continents increase their distance from each other. This is the concept of "sea-floor spreading," so vital for a modern interpretation of plate tectonics. The rate of plate movement is estimated to average 1 to 2 centimeters

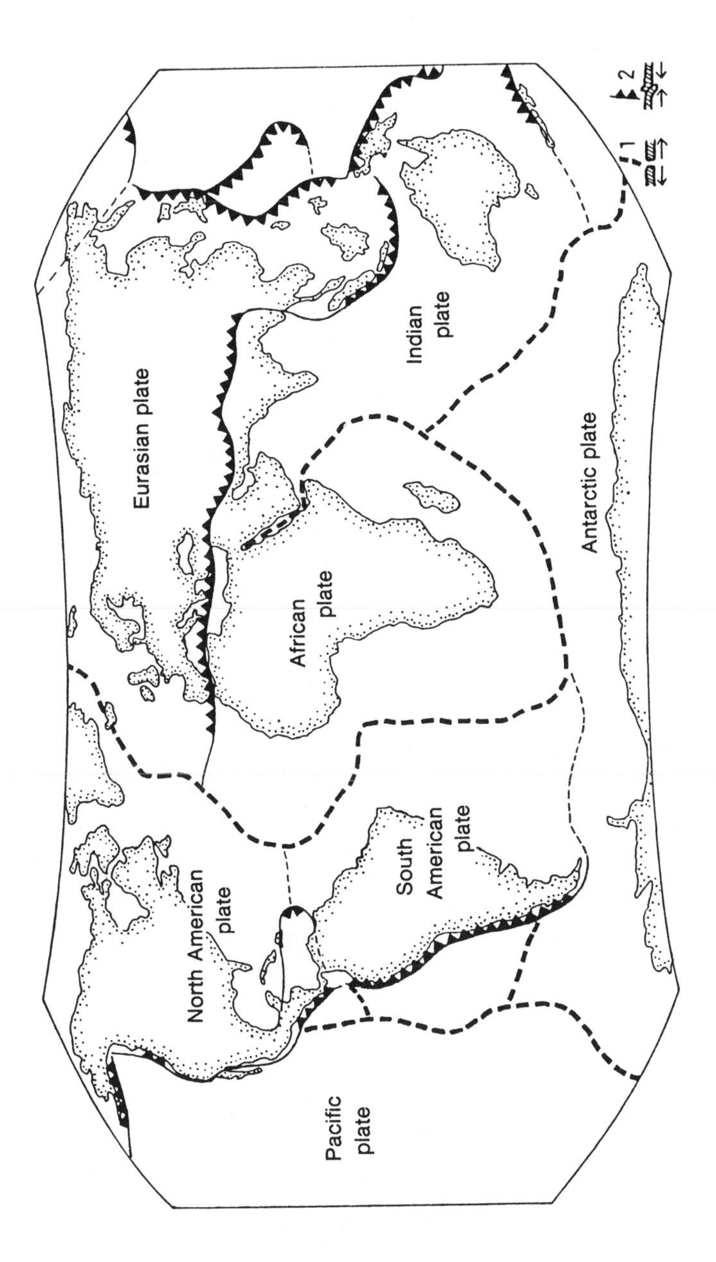

The earth's great "plates." Dashed lines, zones of separation and magma influx (sometimes marked by mid-ocean ridges); notched lines, zones of collision of two plates.

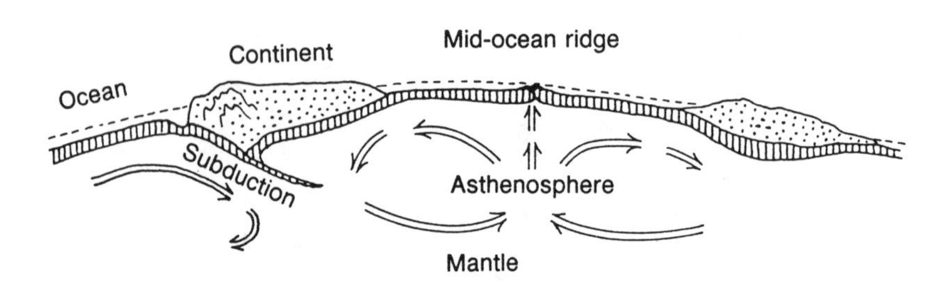

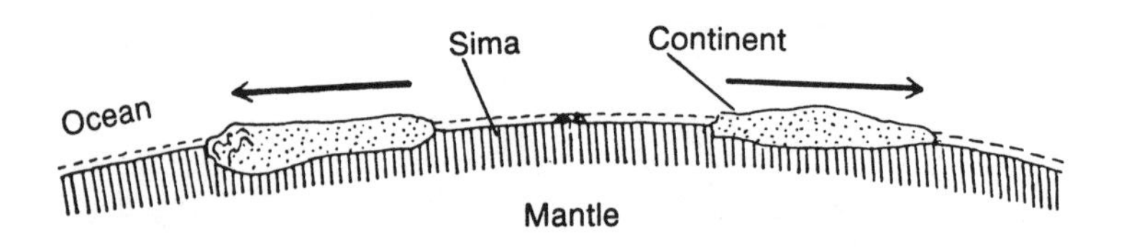

Above: The driving force of continental drift according to plate tectonics: swirling convection currents of the magma in the mantle (in the "asthenosphere"), which drag with them the solid crust and the upper strata of the mantle, including both the continents (dotted area, "sial") and the ocean basins (vertical hatching, "sima"). The drawing also shows a mid-ocean ridge (continuous creation of new ocean floor by the influx of lava), the underthrusting of an oceanic plate beneath a continental plate (subduction), and the consequent formation of a range of folded mountains. Below (for comparison): Wegener's conception. The continents (sial) float on the denser sima. The "pole-fleeing force" was suggested as one possibility for the driving force. The original line of separation of two continents is inactive. The two schematic profiles correspond approximately from left to right to the cross-section; America—Atlantic Ocean (with mid-Atlantic Ridge)—Africa-Europe.

per year. According to Wegener's theory, the ocean floor is old, with new areas coming into view as the continents are displaced. In plate tectonics, however, the ocean floor is very young, youngest at the mid-ocean ridges and older as the distance from the mid-ocean ridges increases.

Pangaea and the formation of mountain ranges

Common to both hypotheses is the assumption that the creation of the earth's great ranges of folded mountains occurs at the leading edge of drifting continents or plates. But once again, plate tectonics has developed models which differ greatly in detail from Wegener's theory. For example, an important concept in plate tectonics holds that the American plates, which are drifting westward, meet the eastward-moving Pacific plate along America's Pacific coastline. Here the Pacific plates slide beneath the American plates ("subduction"), where the submerged plate material melts—an elegant solution to the problem of making room for the new crustal material which is continually being created.

The regional distribution of earthquakes and volcanoes can also be illuminated by concepts of plate tectonics. Here again we should mention Schwinner, who in 1941 connected "deep focus shocks" (with foci at depths as great as 700 kilometers) with large, sloping zones of movement in the earth's crust—"underthrusting from the direction of the ocean, rather than actual overthrusting of the land mass in the direction of the ocean."[7] That is precisely what is assumed today for the collision of oceanic and continental plates. Ten years after Schwinner, the American Hugo Benioff substantiated that assumption with a far greater volume of experimental data, and today these zones of movement are called "Benioff zones." They could also rightfully be called "Schwinner zones."

Like Wegener, the theorists of plate tectonics assumed the existence of a supercontinent in the Carboniferous and Permian, the fragments of which—the present-day continents—later drifted apart.[8] The conclusions which can be drawn about ancient climatic changes are very similar to those of Wegener and are particularly valuable to paleoclimatologists. It should also be noted that Antarctica has yielded much important new data, especially through the work of American and Russian researchers. In Wegener's day, the sixth continent was still almost completely *terra incognita.*

In 1926, when American oil geologists sponsored the New York debate on continental drift, there was no indication that anything of practical value could ever be drawn from Wegener's ideas. Today, that has completely changed. Now geologists search for new mineral deposits not only in the vicinity of established sites but also on the now-distant continents which once abutted them. They are also interpreting the formation of such deposits within the larger framework of plate tectonics.[9] To give just one example, the Austrian geologist Walther E. Petrascheck discussed "Greenland's Mineral Deposits in the Light of Continental Drift" on the basis of ore beds located far from Greenland—in Norway, Scotland, and eastern Canada.[10] Quite aptly, a Viennese newspaper gave an article on Petrascheck's theories the headline "Searching for Ore without a Pick-Axe."

The plate tectonics revolution

With results like these to its credit, it is easy to understand why plate tectonics has been received with such enthusiasm—quite unlike the reception accorded Wegener's theory of continental drift. His hypothesis, of course, had been rejected primarily because some of its proofs were flawed and also because scarcely any direct physical evidence for drifting existed at the time

(Facing page) The drift of the continental blocks since the Paleozoic, according to the theory of plate tectonics. Pangaea splits apart: the position of the continents then changes with respect to the pole and the equator and with respect to climatic zones: arrows indicate directions of movement. Wegener's reconstructions of fifty years earlier are very similar.

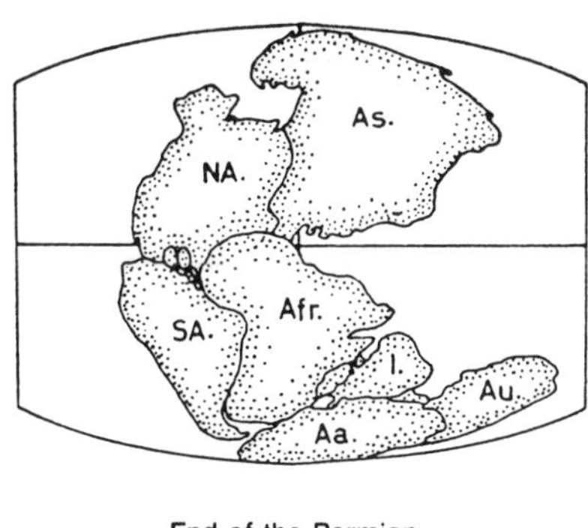

End of the Permian

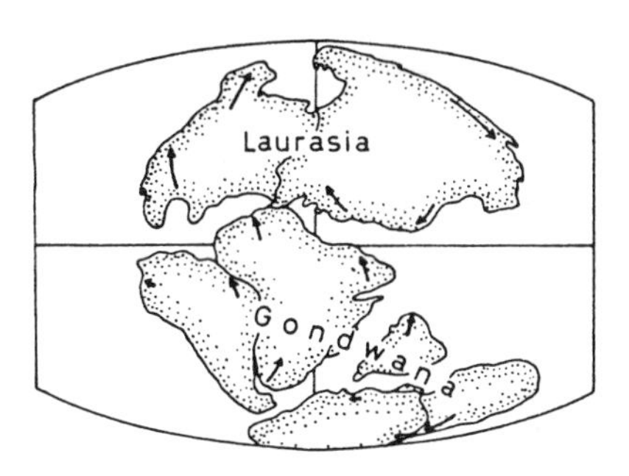

End of the Triassic

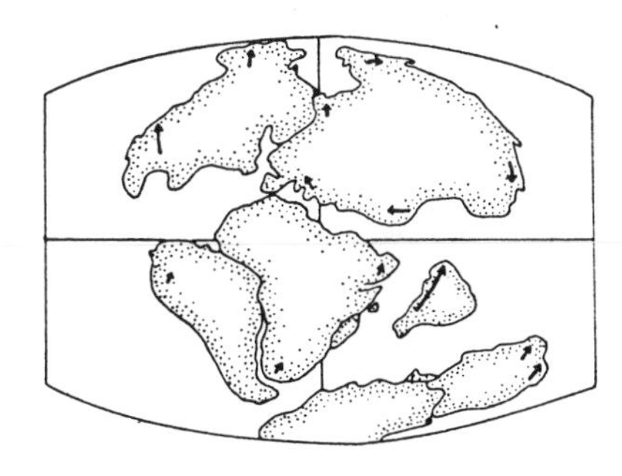

End of the Jurassic

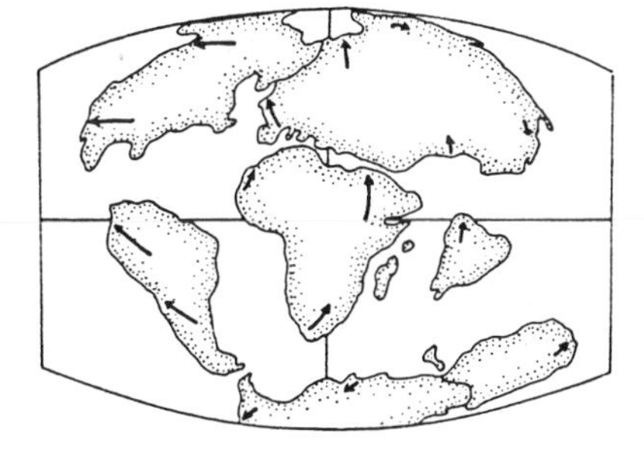

End of the Cretaceous

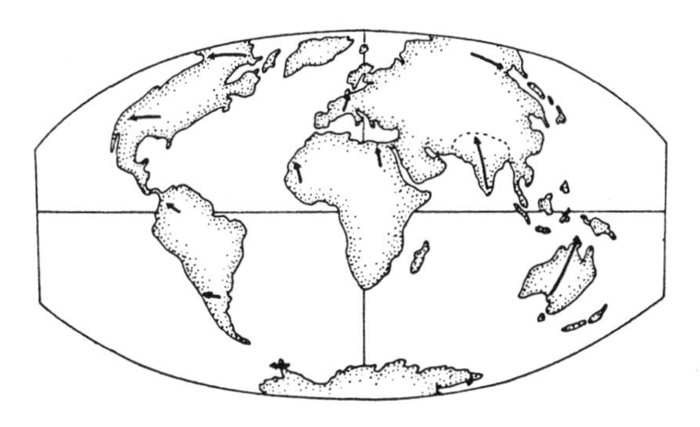

Present

Wegener presented it. Wegener's hypothesis, however, had provided a magnificent, new total view of the earth's development, even if, for many people, it was only the "dream of a great poet."[11] With the advent of plate tectonics, Wegener's idea gained a much more solid foundation, which fascinated even more skeptical scientists, to say nothing of less critical observers.

Plate tectonics has even become the subject of demographic research. In 1977, 128 members of the Geological Society of America and 87 members of the American Association of Petroleum Geologists were surveyed to determine when they had accepted the hypothesis of plate tectonics.[12] At that time 87 percent of those surveyed said they accepted the continental drift hypothesis and plate tectonic theory. Only 22 percent said they had accepted it before 1961. Since the Geological Society of America alone numbers over 10,000 members, the results may not be representative. Nevertheless, only one of those polled thought that the hypothesis would be laid to rest in ten years.

The number of publications on plate tectonics, both important and unimportant, has long since numbered in the thousands, especially in English-language periodicals. One need only page through the volumes of the last twenty years of *Nature* and *Science*, the most widely read journals in the natural sciences, or the publications of the Geological Society of America. Sometimes the papers report on local phenomena, which are then rather arbitrarily linked with plate tectonics, lending some truth to the criticism that the theory is spreading like an "epidemic." And the question, posed half in jest, by Dietz and Holden in 1973,[13] also makes one stop and consider whether we need to speak of a "new orthodoxy" (a way of thinking which—according to the dictionary—tends toward "rigidity and intolerance").

Numerous difficulties arise when one applies concepts of plate tectonics to such well-researched and complex geological formations as the Alps or even older mountain ranges or to the earlier history of the earth in general (*before* "Pangaea"). A few European scientists, in particular, have voiced such criticism.[14] And yet, we must not forget that nature does not always work according to only one plan, or to a simple plan. Discrepancies between the over-simplified clarity of the schematic drawings in textbook discussions of plate tectonics and what one actually observes in the field do not necessarily mean that the concept of plate tectonics is fundamentally in error.

In any event, there are many opponents of the theory of plate tectonics who nonetheless accept the possibility of continental drift. It is not so much the basic idea of the displacement of continents that raises doubts but rather the many unexplained details of the processes occuring both on and below the surface. The convection currents presumed to operate in the mantle change location and direction and vary in intensity. Thus their effects on the earth's surface also vary: the velocity of drift, the time and place of mountain range formation, are obviously complex. It is assumed, for example, that oceans which today are expanding, like the Atlantic, might once have been subject to opposite forces. There are thus many possibilities for changes in the face of the earth in the course of its history. Most of the details are still not well understood.

Both the true believers in the drift theory and the uncompromising anti-driftists are guilty of the same error as Wegener and his opponents before them: from the many geological observations available, the so-called facts, they have all too often selected those which support their position, while leaving out others which argue against it. That is only human and is not necessarily intentional; sometimes, it is probably a matter of convenience. Moreover, the same thing has also been characteristic of other great scientific debates.

A fascinating total view

In spite of all the criticisms of plate tectonics, it must be said that none of the previous tectonic theories—from the wrinkled apple of the venerable contraction theory to the expanding earth of Otto Hilgenberg, S. Warren Carey, and Pascual Jordan[15]—offers such a fascinating total view. Scarcely any other concept in the history of geology in the last 200 years—Darwin's theory of natural selection excepted—has caused as much excitement as plate tectonics, the modern form of the theory of continental drift. Of course, it is still true, as it was in Wegener's time, that many observations remain enigmatic. Warner Zeil wrote, "Today's model of plate tectonics—as imposing as it is—provokes many more questions than it can answer."[16] Many inconsistencies must still be resolved: more through geological observation than through arguments from physics, in Vine's opinion.[17] At the close of his 1977 review of the continental drift debate, he quoted Sir Arthur Eddington,[18] the well-known British astronomer: "The time has gone by when the physicist prescribed dictatorially what theories the geologists might be permitted to consider." Lester C. King made the same point in a lecture in London in 1957: "The driftist is no more obliged to adduce a mechanism to prove the fact of drift than the user of an electric appliance is obliged to define the nature and mechanism of electricity."[19]

But we can go back even further, to Galileo, 300 years ago: "Experience and sensory perception take precedence over all kinds of speculation, no matter how well founded they may be."[20] However, the "experiences" of geologists only occasionally yield more or less certain facts. Generally, they do not bear comparison with the measurements of which the "exact" natural sciences are capable.

8

RECOGNITION AND FAME

He surely qualifies for a niche in the pantheon of great scientists.

—ANTHONY HALLAM[1]

Recognition and Fame

Fame and honor are often connected, although honor is usually expressed only through rather superficial forms of recognition for some achievement or other. Wegener the scientist was little honored during his lifetime, and when he was honored, it was as a polar explorer and not as the originator of the theory of continental drift. That was true as early as 1908, when the King of Denmark awarded the Silver Medal of Service to the young Wegener upon his return from his first Greenland expedition. In 1913, after completing his second expedition, he received a second Danish medal, the Knight's Cross of the Order of Danebrog. In the same year, both Wegener and J. P. Koch were awarded the Carl Ritter Medal by the Gesellschaft für Erdkunde in Berlin; and in 1923 Wegener received the Kirchpauer Medal from the Geographischen Gesellschaft Hamburg "in recognition of his important achievements in northern Greenland." As a professor in Graz, Wegener became a corresponding member of the Vienna Academy of Sciences.

After his death, the city of Graz honored him by renaming the Blumengasse, where he had lived, as Wegenergasse; and today a bronze bust of Wegener, the polar explorer, by Wilhelm Rex, stands in the main university building next to those of two Nobel laureates. The bust was formally given to the university by the German Minister of State Friedrich Schmidt-Ott on behalf of the Notgemeinschaft der Deutschen Wissenschaft. In 1961, Austrian alpinists placed a memorial tablet, donated by the University of Graz, on the north side of Kamarujuk Fjord—the site of the West Station for the 1930 expedition—in western

Greenland; and the large peninsula north of the fjord was named Alfred Wegener Peninsula by the Danes. There is also an important German institute for polar research named for him in Bremerhaven, the *Alfred-Wegener-Institut für Polarforschung*. In Hamburg, the road leading to the city's youth hostel is named Alfred-Wegener-Weg (see page 138); the hilltop site is the former location of the German Marine Observatory and is near the present-day Hydrologic Institute into which the Marine Observatory was incorporated. Also in Hamburg, there are streets named for Köppen and Georgi. In Berlin, a secondary school is named for Alfred Wegener. Recently, memorials for Wegener have been erected in Graz, Marburg, and East and West Berlin.

His memory has also been honored in the German Democratic Republic. In the old "director's house" in Zechlinerhütte (also see page 12), which Wegener's parents used as a summer home, two rooms have been converted into a museum, a "Wegener memorial" for Alfred and Kurt Wegener, with memorabilia and photographs, much of them contributed by Else Wegener. However, the museum's range is narrow, with most of the exhibits relating only to Wegener the Greenland explorer, rather than to his revolutionary scientific importance. Nor does the museum attract many visitors other than summer tourists vacationing in the out-of-the-way village.

True fame? According to the dictionary, fame is "the public estimation lasting beyond one's lifetime." Wegener's name was known beyond Europe during his lifetime; his ideas about continental drift were respected by many; there were occasional conferences devoted exclusively to continental drift. But it is difficult to say whether on that basis he could be considered famous, especially since, as we have seen, there was a period when his ideas were no longer being discussed, and since he was known less for his pioneering scientific achievements than for his Greenland expeditions. Although the German Meteor-

Memorial tablet for Alfred Wegener, which was placed near Kamarujuk Fjord in Greenland, donated by the Karl Franzens University of Graz. The bronze tablet was executed by Wolfgang Skala; the dog sled and Wegener's Eskimo cap evoke the Greenland explorer.

Street in Hamburg renamed in the 1970's for Alfred Wegener.

ological Society has been awarding an Alfred Wegener Medal for the past several years, the honor inherent in the award's name is intended more for Wegener the meteorologist than for the scientist who formulated the hypothesis of continental drift.

Today, plate tectonics, which developed from that hypothesis, is a commonplace among earth scientists; it has become a "famous" hypothesis. To be sure, Wegener's name is seldom mentioned in the countless papers on the topic, but of course the same is true for Copernicus, Newton, and so on, in other fields. But Wegener has certainly not been forgotten. Many textbooks call him the true father of continental drift. "The name most commonly and correctly associated with the hypothesis of continental drift is that of the German meteorologist Alfred Wegener," wrote F. J. Vine in the introduction to his detailed 1977 review article in *Nature*.[2] Thus we are justified today in calling Wegener a famous scientist. His importance was only recognized after his death, and thus fame came to him only posthumously. But surely that is of greater value than the often transitory fame enjoyed during one's lifetime. Furthermore, as Anthony Hallam stated, "It is interesting to note that, although the propounder of continental drift was German, major post-war developments in the subject have been almost an Anglo-American monopoly."[3]

There is also another curious aspect to the history of the theory of continental drift. Wegener's innovative conception that the continents are moving is still a vital part of earth science today. The idea came to him when he was a young man and occupied him during his entire lifetime; and he tried again and again to support and expand the concept with new observations and data. Unfortunately, he did not succeed very well. While many of his arguments were indeed correct, others were soon refuted. Today's advocates of his drift hypothesis are able to substantiate the theory with different arguments and with greater success.[4] One could say that Wegener suggested a revolutionary solution to an important problem of the earth's history, thereby providing an extraordinary stimulus to scientific thought, but that his "working out" of the problem was inadequate and only partly correct. Altogether in the same vein was Albert Einstein's judgment of a paper by his highly esteemed colleague Max Planck, which contained an error: "The solution is correct, but the argument is false." To what extent the argumentation in favor of plate tectonics is sound is, of course, also uncertain.

However, that does not really diminish the fundamental importance of Wegener's hypothesis. In any event, today no one could say "the theory of continental drift is a fairy tale" as Bailey Willis did in 1944. On the other hand, it seems exaggerated to regard Wegener as a latter-day Copernicus, as has sometimes been done. As early as 1922, in an unsigned review of the second edition of *The Origin of Continents and Oceans* which appeared in *Nature*,[5] the revolutionary concept of drifting continents was compared to the Copernican revolution. One of the most prominent co-founders of plate tectonics, the Canadian J. Tuzo Wilson, took up the comparison again in 1968:

> In our day it would appear that what earth science needs more than fresh data, better instrumentation, or new techniques is a simple change from our present belief that the structure of earth is static to the new concept that it has long been mobile. This is parallel and similar to the Copernican revolution and should perhaps be called the Wegnerian revolution, from its chief advocate.[6]

In addition, the legendary comment of Copernicus's unfortunate advocate, Galileo, "Eppur si muove" (nevertheless, it does move), also fits Wegener. With the statement that the earth moves around the sun as with the statement that the continents are moving, concepts which had stood firm for centuries were overturned. And yet, the intellectual dimensions are quite different. Copernicus changed not only assumptions of physics and astronomy but also revolutionized the religious and philosophical thought of his time, which is not true of Wegener's hypothesis. And a second difference: Copernicus's heliocentric world view was later confirmed as fact. In the case of Wegener's original idea as well as the case of plate tectonics—the modern version of the continental drift hypothesis—well-substantiated proof is still lacking.

However, no other German earth scientist has provided such a sustained stimulus to research both within and outside of his own country as Alfred Wegener, not even his Berlin compatriot Alexander von Humboldt, even though Humboldt's name enjoyed and still enjoys far greater renown throughout the world. In a memorial address at the Berlin Academy of Sciences in

1888, long after Humboldt's death, Emil Du Bois-Reymond characterized Humboldt's influence very aptly:

> Let us not delude ourselves: the only member of the mathematics-physics section of the Academy who has been honored in Berlin with a public monument, Alexander von Humboldt, owes this honor not to the scientific achievements which keep his memory alive in this hall, but to the grand memories connected with his name, his gripping depictions of the natural world, the glow of his enthusiasm for the true and the beautiful, and the unparalled international renown he enjoyed at the culmination of his long and happy career.[7]

In Wegener's case it was the reverse. His career was neither brilliant nor long, nor did it end happily. To be sure, his fame—very modest in comparison with Humboldt's—was also due at first to the perilous journeys he undertook; but his true importance to science is found elsewhere altogether, in a new revolutionary hypothesis, whose influence is strongly felt today.

Beyond dispute and untouched by questions of recognition and fame, is the memory of Wegener's character, which his friend and colleague Hans Benndorf described at the conclusion of his obituary note: "His was a character of flawless purity, straightforward simplicity, and unusual modesty. But at the same time he was a man of action, whose iron will and unflagging energy led him to extraordinary achievements, even as he risked his life in the pursuit of an ideal goal."

He certainly numbers among the great natural scientists.

9

MEMORIES OF ALFRED WEGENER

BY JOHANNES GEORGI

MEMORIES OF ALFRED WEGENER

Memories of an important man will always contain a subjective element, the overall picture of him emerging only from the sum of many individual glimpses from different angles. Since my meetings with Wegener were mainly in the fields of meteorology and Greenland exploration it is advisable to turn not only to my personal memories of the years 1910 to 1930 but also to other personal sources, above all to his biographer, Professor H. Benndorf,[1] who was his friend and colleague at the Physical Institute of the University of Graz (Austria) from 1924 to 1930. I am also grateful to Professor W. Wundt of the University of Freiburg i.B., a fellow student of Wegener's in Berlin from 1903 to 1905, for letters relating to that period. The two became acquainted during Professor von Bezold's meteorology seminars and practicals and, having many interests in common, became close friends. Wundt informs me that even at that time (1903) Wegener showed him the route by which he wished to cross the Greenland ice cap at a more northerly latitude than that taken by Fridtjof Nansen in 1888; and at the same time Wegener pointed out to his friend the apparent congruence of the western and eastern Atlantic coasts—an idea which, according to S. K. Runcorn,[2] has continuously fired the imagination of eminent explorers such as Alexander von Humboldt[3] since the time of Francis Bacon (1561–1626).

We know that Wegener reexamined this problem in 1910, that early in 1912 he gave his first public lecture about it to the Geological Association, and in the same year brought out his first great publication on the origin of the continents.[4] We also know how strangely and unexpectedly this field of work came into contact with Wegener's Greenland explorations in that measurements of longitude on Sabine Island (the larger of the Pendulum Islands, East Greenland 74·5°N) since the time of Edward Sabine in 1823 seemed to point to a continuous westward drift of Greenland and thus to give a proof of the theory of the horizontal displacement of continental landmasses (a supposition that happened to prove untrue).[5]

The "Grand Old Man of Meteorology," Professor W. Köppen of Hamburg, with whom Wegener had close contact from 1910 onwards and whose daughter became his wife, adds[6] a not unimportant detail to the story of the development of the displacement theory: Wegener was led to a serious analysis of the problem by a compendium which "fell into his hands" one day in 1911, from which he learned of the great similarity in the older faunae of South America and West Africa: this led him to study all the available literature on the subject. Köppen, who incidentally[7] had warned Wegener against spending too much time in allied fields in order to be able to devote all his energy to his meteorological research, remarks that probably many a scientist when looking at the map of the world had already wondered at the similarity of the Atlantic coasts. "But now this similarity had been noticed by an expert geophysicist, a brilliant man of unbounding energy, who would spare no pains in following up the matter and gaining any facts from other fields of science that might seem to have a bearing on the question."

Köppen also foresaw the drawback from which Wegener was later to suffer so much:

> To work at subjects which fall outside the traditionally defined bounds of a science naturally exposes one to being regarded with mistrust by some, if not all, of those concerned, and being considered an "outsider." The question of the displacement of continents involves geodesists, geophysicists, geologists, palaeontol-

ogists, animal- and plant-geographers, palaeoclimatologists, and geographers, and only by consideration of all these various branches of science, as far as is humanly possible, can the question be resolved.

Let us consider for a moment how strongly these two interests had gripped Wegener even during his student days. His friend, Wundt, recalls that he was already testing his powers of endurance by long, exhausting walks, and by days of skating through the winter-bound Spreewald, getting himself in training for the Arctic. In 1901 at Innsbruck during the summer holidays he undertook with his brother, Kurt, very ambitious alpine climbs that were not entirely without danger; and in the winter of 1903–04, when visiting a friend, who was a meteorological observer on the Brocken, the highest mountain in the Harz, he turned seriously and enthusiastically to skiing. From all of which there emerges a sense of purpose and determination that is encountered all too rarely and which always heralds great achievements.

There was the same aim in his practical study of the new technique of aerology, using kites and captive balloons, which had just been developed by Rotch, Teisserenc de Bort, Köppen, and Assmann at the Royal Prussian Aeronautical Observatory at Lindenberg, to master the then most modern and most difficult meteorological methods in order to make use of them later in his own field work. In the service of the Observatory he and his brother took part in 1906 in the Gordon Bennett Contest for Free Balloons. With an unbroken flight of fifty-two hours they easily broke the world record of thirty-five hours.

Such careful preparations have probably never been more handsomely rewarded, for in 1906 Wegener was allowed to accompany, as official meteorologist, the Danish "Danmark" Expedition led by Mylius-Erichsen to northeast Greenland, where he spent two winters at more than 77°N. From there he made long and difficult journeys with dog sledges south to 74°, and north to almost 81°; and with manhauled sledges he went up to the edge of the ice cap. During that time he was able to put into practice all his physical and intellectual abilities, his scientific knowledge, and his practical skill for the exploration of the upper layers of the atmosphere in the Arctic, for meteorological investigations of all kinds, for astronomy, meteorological optics, and glaciology, as is shown by publications which have not, even today, been superseded.[8] Equally he was able to stand his ground in fitness and endurance with the experienced polar explorers in the expedition.[9]

After a safe return, Wegener settled down in Marburg, working on his abundant Greenland material and giving lectures and demonstrations in astronomy and meteorology. When I arrived there at Easter 1910 for the finals of my studies in physics I found a notice on the board of the Physical Institute, in a clear and attractive handwriting, announcing that *Privatdozent* Dr. Wegener was to give some meteorological lectures and demonstrations. In the small observatory of this famous old Institute the three or four students who were interested met the hearty, fresh-faced young man of medium height who was to be their tutor and who quickly won their hearts by the firm yet at the same time modest and reserved manner in which he immediately introduced them to the fundamentals of the subject. Only here and there could we catch a glimpse of the lion behind the lamb-like manner, as when he criticized certain precautionary measures necessary for working under extreme climatic conditions, which were not then practiced.

His lecture on "The Thermodynamics of the Atmosphere" was remarkable for the ease of his delivery, which was in complete contrast to the difficulty of the subject. Numerous examples were taken from his recent observations in Greenland; and here for the first time the attempt was made to relate the bulk of measurements from the free atmosphere during the last dozen years to general physical rules for the explanation of the manifold phenomena, such as the different atmospheric layers (only eight years had passed since the discovery of the stratosphere!) and the various types of cloud formation. Whoever in those days had the opportunity of following the lectures and practical work of famous scholars would have had to admit that Dr. Wegener's lecture bore no professorial stamp at all. On the contrary, the tutor came down to the level of his audience and developed with them the theme which he had just set down in an epoch-making book.[10] It is true that the final result still had to be formulated mathematically, but neither before nor since have I had the experience of hearing a tutor state quite simply: "This deduction is not mine; you will find it in the physics textbook by. . .on page. . . ." This little anecdote points to the fact that Wegener had no special gift for mathematics. He had already remarked to Wundt, to whom we are indebted for many illuminating stories of Wegener's early life, that, although he had made considerable advances in astronomy, he had no great inclination for it, and considered that for any great achievements in this direction a special mathematical gift was necessary, which he himself lacked. Frau Else Wegener quotes from a letter[7] from her former fiancé to her father, in which, after first calling his attention to a similar concept of G. H. Darwin's in his standard work on tides,[11] he continues:

> I myself hold the crass and probably exaggerated point of view that such mathematical treatises as I cannot understand (i.e., in which I can no longer follow the gist of the thing—for one can often follow the gist without working through the formulae) are wrong or do not make sense. When one cannot follow the printed or written word one should not always put the blame on oneself. When logic is lacking, one can still usually fill out a few lines with formulae."

If his talents here show a gap, he always compensated in two ways: firstly, he always took the greatest pains, even in his most specialized work, to be as intelligible as possible and not to write only for his fellow experts, and secondly, an outstanding trait of his character was his frankness even towards his students. He had an unusually high degree of integrity in such a natural and unpretentious way that one had the impression that he was exempt from the common human temptation of occasionally making oneself appear a little more important than one actually is.

I am sure that young people in particular feel this immediately; and the simplicity in his lectures and demonstrations, obviously based on experience and achievement, always won him the hearts of his audience. At the end of the lectures in Marburg he used to bring out a number of photographs to illustrate the subject he had been discussing, usually cloud formations, the refraction of light in the layers near the ground, air-optical phenomena produced by the reflection and refraction of light in ice crystals, or photographs of the formation, movement and transformation of sea mists; mainly photos that we were the first to be shown. These pictures were discussed as examples of the subject just dealt with in the lecture. It was also an innovation for us students to see that senior academics, such as the assistants at the Physical Institute, and in particular Professor K. Stuchtey, who carried out several free-balloon measuring flights together with Wegener and later remained his loyal helper, did not consider it beneath their dignity to come and listen to these lectures by a young private tutor. How highly Wegener valued photography as a method of research he himself mentioned in a letter (an understatement, of course): "On Koch's expedition (i.e., on Koch-Wegener's Greenland Expedition 1912–13) I did practically nothing but take photos: clouds, ice, micro-photos, aurora, mirages, color photos (using the Lumière system), neutral spots, always using different cameras and plates. Even on the "Danmark" Expedition, as well as the normal photography, I used the Miethe Three-Color Method of photography which, of course, because of technical difficulties, could only rarely be shown in lectures."

The winning simplicity of his nature, which was seen for the first time in Marburg, remained with Wegener all his life, even long after he had become famous. This we see from a reminiscence of the late twenties from his friend and colleague Professor Benndorf[1]:

> In the afternoons he was always a guest to tea in the Institute. He would regale us then with stories of his travels, and the students who did not know much about his experiences listened eagerly and with the greatest attention. That our young academics who are so sports-minded and, in Graz, especially keen on skiing, were highly impressed by his feats goes without saying. Most of them also knew that he was already a famous scholar. But one would never have guessed it by looking at him, and it was the simple and unpretentious manner that he had of talking on an equal footing to even the youngest student that took the hearts of the young people by storm. I think that they would have gone through fire for Wegener; and if anyone had dared to doubt the theory of continental drift, they would not have hesitated to use their fists as arguments.

It was a hard blow to his students and assistants when in early 1912 Wegener once more gave up his academic work in Marburg to make another trip to Greenland—and Greenland was heartily cursed, especially by me, as I had become fired with enthusiasm and wanted to stay with this teacher and devote myself entirely to meteorology, which was then possible at only a few universities. But for Wegener himself, the new expedition with his friend, of 1906–08, the Danish Captain J. P. Koch, spending the winter for the first time ever on a high Arctic glacier and the subsequent crossing of the ice cap for a distance of 1,200 kilometers, was the completion of his work on Greenland's climatology and glaciology, begun on the "Danmark" expedition. At the same time this undertaking made the highest of claims on his physical capabilities so that all in all it was probably his greatest achievement as a man—the account of this journey is and will remain a classic of the exploration of Greenland.[12]

Scarcely had Wegener, now newly married, set foot again in Marburg than he was caught up into the maelstrom of World War I and was twice wounded. In 1919 he took over, as W. Köppen's successor, the Meteorological Research Department of the Deutsche Seewarte (German Marine Observatory) in Hamburg which was to be completed by the building of a modern meteorological experimental station on the grounds of Köppen's historic Balloon Station at Grossborstel, north of Hamburg. As I too was to work there, my personal relationship of many years standing with Wegener was once again renewed. If the new building, planned to the last detail and unique of its kind, fell a victim to inflation, nevertheless a period of highly interesting and inspired collaboration began.

During the preliminary discussions for the new experimental work that was planned I observed how cautiously Wegener made use of his intellectual superiority. We were walking along the corridors of the Marine Observatory one day while Wegener talked about various experimental equipment and also wanted to hear my suggestions. Although a meteorologist, I happened to have read about some hydrodynamic experiments with pulsating balls in water, which had been described in 1876 by the father of the famous meteorologist Wilhelm Bjerknes. Wegener listened with interest, and I was flattering myself that I had told him about something new when, without any unkind intent on his part, he showed in the course of conversation that I had not mentioned all the facts of the matter—in short, that he knew far more about those old and rather off-track experiments than I did. That was not the only occasion during the course of the next ten years that I came away red in the face after conversations with him, shamed by his more extensive knowledge and at the same time by his kindheartedness.

With the technical facilities now at his disposal in Grossborstel he was able to investigate various instrumental problems such as the failure of clockwork meteorological and aerological recording instruments when used in a polar climate or in the upper atmosphere. He also designed a new and efficient balloon theodolite for following pilot or other balloons from ships up to great heights, an instrument useful for decades afterwards. This problem he solved in a way characteristic of him, by combining a normal pilot-balloon theodolite with an ordinary ship's sextant. I also recall a big rotating iron drum used for experiments to show the effect of the rotation of the earth on fluid motion.

But while, several times a week, he had to cope with official correspondence in the Marine Observatory, which was to his great distaste and which he rightly regarded as a waste of his time and energy, he could also, in his primitive workroom in the Meteorological Experimental Station, at last devote himself to the theory of continental drift.*

The working out of this was now progressing well. After the end of the war, with the renewal of contacts with the rest

*Imbued as we are with Wegener's ideas, we can no longer appreciate the revolutionary effect they had on scientists. As an example of his influence, let us look at the famous lecture delivered by Alfred Penck in 1885. Still only a young man, this leading glacial geologist who had at his command a vast collection of geological and paleontological data, came to the conclusion that both poles of the earth, then regarded as invariable, had been the centers of extensive biological development (*Verhandlung 5.deutsch. Geographentag*, Hamburg 1885, pp. 5–50, 174–196). At that time this conclusion must have been regarded as both theoretical and abstruse, but thirty years later Wegener's theory supplied the key to the solution of Penck's astonishing enigma.

of the world, there came not only news from colleagues in various faculties about new findings for or against Wegener's theory, but also these experts from all over the world came in person to visit the modest wooden huts in Grossborstel or the nearby Köppen-Wegener house. At that time one could regard Grossborstel as the Mecca of geophysicists and ecologists interested in this problem just as, twenty years earlier, because of Köppen, it had been the Mecca for the new science of aerology. The most important result of the continental drift theory was its application to palaeoclimatology by Köppen and Wegener working in collaboration.[13,14]

For those of us who were Wegener's scientific colleagues in Grossborstel and who, of course, were also on a very friendly footing with our chief, there were exciting days when new support for his theory presented itself, and depressing ones when he had to argue with his opponents or even defend himself against apparent misunderstandings. We had the good fortune during this time to meet many famous scholars; in particular I remember a visit from Wegener's former Greenland colleague J. P. Koch, then a colonel on the Danish General Staff, a man who in appearance and manner was just as we had always imagined the Vikings of the Icelandic sagas. Their discussions, about the meteorological and glaciological problems in Greenland then being tackled and about those that still remained, increased my appetite for scientific work in the polar regions, which had already been whetted in Marburg.

A source of great pleasure were the informal invitations to the Köppen-Wegener home. Our houses were only a few yards apart, and our children used to play in each other's gardens. Köppen himself was a man of uncommonly wide interests and a most stimulating companion, especially on the social-ethical side. His neighbor, a highly respected schoolmaster, and a convinced and active pacifist, was very close to him and to his household. Both families were also friendly with a well-known Hamburg lawyer, who represented in word and print the philosophical preachings of W. Ostwald. There is no doubt that Wegener also was sympathetically inclined towards these views as we, his later colleagues on the 1929–30 expedition, were often able to observe. I was very impressed, however, on our occasional visits by Wegener's taking his leave of the little company, to attend to urgent work in his study, usually correcting proofs. Without such economy of his time his astonishing productivity would not have been possible. We were later to witness, on the journey out to start the preparatory expedition to Greenland in 1929, the iron devotion to duty which he showed by correcting proofs on board ship, even during heavy seas which claimed the rest of us as seasick victims. The voluminous proofs for the great two-volume work on the expedition to Greenland by himself and Koch in 1912–13,[15] were dealt with as carefully as at his desk at home.

It may be mentioned in these personal reminiscences how very much Wegener's colleagues regretted the fact that this great scholar, predestined for research and teaching, could not get a regular professorship at one of the many universities or technical high schools in Germany. One heard time and again that he had been turned down for a certain chair because he was interested also, and perhaps to a greater degree, in matters that lay outside its terms of reference—as if such a man would not have been worthy of any chair in the wide realm of world science. So our universities had to look on while in 1924 a regular professorship in meteorology and geophysics was created especially for him at the University of Graz in Austria. This post met his approval, partly because of the number of colleagues there with like interests and partly because it freed him from the burden of administrative duties, to the advantage of the whole world of geophysics.

For myself a new period of even closer association with Wegener began in 1928. In 1926 and 1927 I had been doing aerological work for the first time in the far northwestern point of Iceland and had discovered there the high storms known today as the "jet stream." I was planning an aerological winter base on the Greenland ice cap for further research and it was only natural that I should inform my teacher and former chief of this plan, as he was the greatest expert on Greenland, and ask him for his opinion and support. As his reply shows, this request seems to have caused an inner revolution in him: ". . .You mention also the question of a station on the inner ice cap. That is an old plan of Freuchen's, Koch's, and mine! If only the (First World) War had not happened, it would have been carried out long ago. But meanwhile Freuchen has lost a leg, Koch is in hospital, and I too have had my own trouble and am no longer a young man. I intend to write an article in the first number of *Arktis* about the conditions of work and the scientific problems of such a station. Here only a few things. . . ."[16] There follows in the letter an outline program which enlarged my own limited conception of measuring high winds into a complete geophysical program, with a winter base in the middle of Greenland, such as I had envisaged, and in addition a similar scientific station in West Greenland. He added almost prophetically, with double underlining which was rare for him: "Both stations must remain there for *two* winters, otherwise they cannot fulfill their purpose." However, this could not be carried out because of his tragic death on his great 1930–1931 expedition. The letter ends: "You will see from this how much your plan interests me. I should like it very much if you would keep me informed about its further development."

I was soon able to give him positive news because of the favorable attitude of the German Science Emergency Committee [*Notgemeinschaft der Deutschen Wissenschaft*] in Berlin and especially of its president, Minister of State Dr. F. Schmidt-Ott. I was entirely in agreement when the organization of the

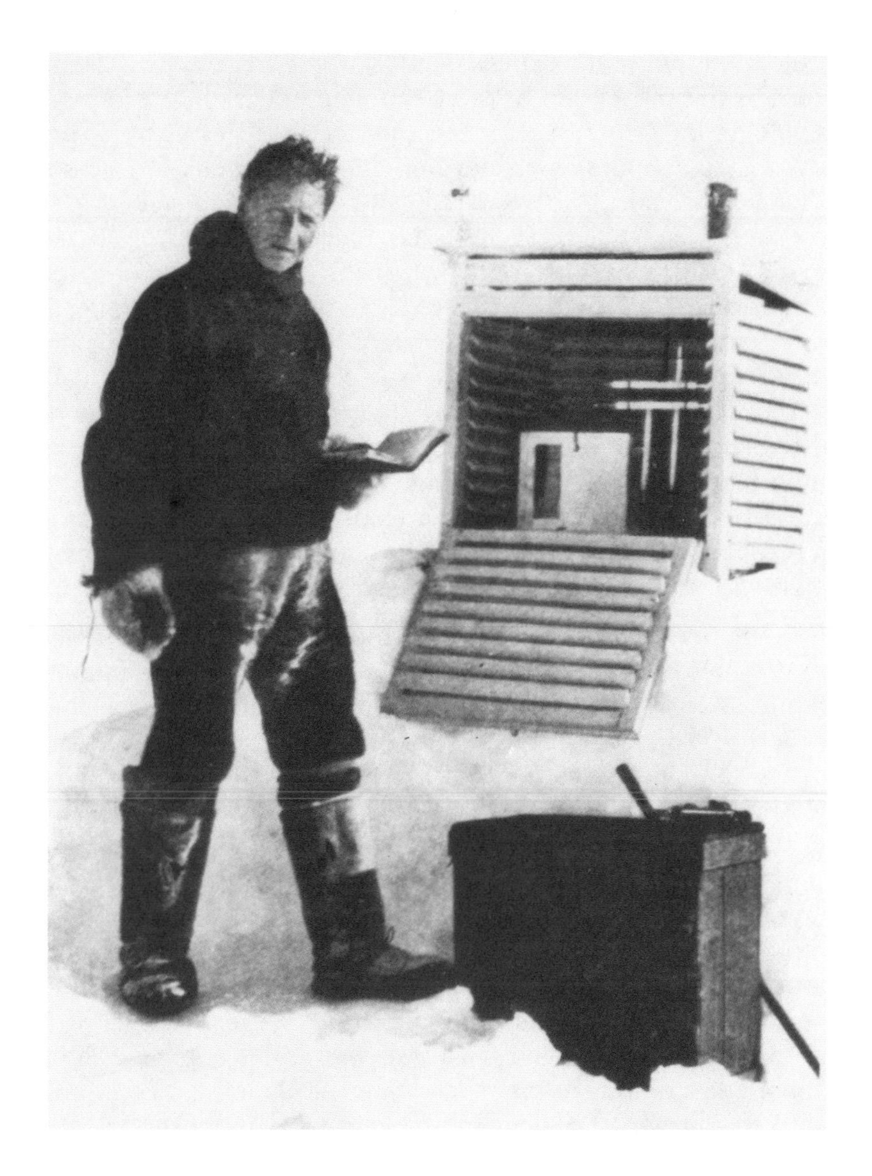

Johannes Georgi taking a weather observation. On the top of the shelter are the wind gauge and sunshine recorder; inside are the thermohydrograph (left), normal thermometer, and low-temperature thermometer.

project was put in the experienced hands of Wegener and when, later, at the instigation of the Göttingen geographer Professor W. Meinardus, the task of measuring ice thicknesses by means of a seismic method newly developed in Göttingen by Professor Wiechert was also incorporated. This three-fold origin of the last and greatest of Wegener's Greenland expeditions is described in his expedition plan[17] on which the preparation and execution of the expedition were still based even after his death in November 1930. It is an important document in the history of polar exploration, especially as it combines the "classical" methods of travel by hand- or dog-sledges with modern methods of transport. It is to be regretted that this work, which was so important as an intermediate stage in geophysical polar exploration between the two world wars and also so characteristic of Wegener's thought and work, has remained almost unknown, for it was not published in the big scientific report.[18]

The personal relationship resumed by our correspondence in 1927 to 1928 continued with Wegener's visit to Hamburg in July 1928, followed by further visits there and also meetings in Berlin until the time came for the departure of the expedition. It was characteristic of him that he came to Hamburg in July not only to arrange my release from the Marine Observatory but also to get my wife's approval. He himself was engaged when in 1912 he went to Greenland for the second time and knew how much those who stayed at home suffered from the separation, especially if the expedition could not continue according to the plan. I shall never forget his long and friendly conversation with my wife; how Wegener explained, when naturally she asked what risk was entailed for me, how he hoped to minimize the risk by careful planning, keeping in mind his own experiences there; how what one usually calls bad luck is very often only a result of errors or inadequacies in preparation.[19] This assertion of Wegener's in 1928 is almost word for word the same as that of the eminent, later Greenland explorer, R. A. Hamilton who, as second in command and scientific leader of the British North Greenland Expedition of 1952 to 1954, twice wintered north of 77°.[20] Here may be cited another example of how tireless Wegener was in trying to avoid causing suffering or even misunderstandings. He traveled to see a former colleague for the sole purpose of explaining to him why he judged it better *not* to invite him to take part in his Greenland Expedition.

This is not the place to tell of the hectic rush that accompanied the preparations for the 1929 expedition, when sometimes there were three letters or telegrams a day from Graz to Hamburg and often as many back; it has been described in many books dealing with the expedition, also in that of R. A. Hamilton.[20] The preliminary expedition itself, whose most important responsibility was to search for and prepare an ascent for heavy equipment from the west coast of Greenland up to the ice cap has been described by the leader himself in a most delightful book,[21] and one can read between the lines more about the man and his personality than from the best descriptions of his character by others.

One might ask why Wegener made no provision in his planning of the expedition for investigating the supposed westward drift of Greenland even though an experienced geodesist had joined the main expedition of 1930–31 for the purpose of carrying out latitude and longitude determinations and geodetic levelling into the interior. Technically, it would have been possible to have made really accurate measurements of longitude using radio time comparisons but this scheme was rejected with some regret as the Danes had recently begun making precision longitude measurements in Greenland with a fixed meridian circle at Qôrnoq near Godthaab[5], and our expedition, with the means at its disposal, would never have been able to reach the same degree of accuracy. How preoccupied he was, however, with the western drift of Greenland his colleagues know from conversations during the long reconnaissance marches the party made on glaciers and the ice cap; they knew just how content the actual renaissance of this idea of his would have made him.

The preparation of the great Main Expedition of 1930–31 in the few months between our return from Greenland in late autumn 1929 and the departure of the first ship for Greenland in April 1930 must have been a nightmare for Wegener; negotiations with the Danish government, who valued and supported Wegener's person and project—and supported it in spite of doubts which had arisen about the gravimeter measurements; negotiations with the Emergency Committee for German Science which, because of difficulties in the Reichstag, was not at the critical moment able to supply the funds which had long ago been granted in principle; the search for suitable members for the expedition's three stations and time-wasting negotiations about this with institutes and authorities; the making of lists of requisites for the hundred and one daily needs of the expedition for two years; and the delegation to individual members of the responsibility for certain things such as clothes, provisions, and instruments, wherever Wegener had not taken this over himself. He made many journeys to Berlin, to Copenhagen, even to Helsinki where Finnish propeller-driven sledges were inspected and ordered. In addition to all this he had his lecturing in Graz and also, after Koch's illness, the sole task of completing the great report about the 1912–13 expedition.

The thick folder of letters that I received from him during these months shows Wegener to have been astonishingly unremitting, not only as far as the continual appearance of new problems was concerned, but also in the sphere of human relationships where his equanimity and kindliness were often put to bitter tests. Often during these months of hard and unenjoyable work (out in Greenland, one assumed, physically harder work would be waiting for him, but it would at least be more congenial to him) he may have thought of his experiences as a young and lively member of the earlier expeditions and of his sometimes quite sharp criticism of the leader and his fellow-members on the "Danmark" Expedition. This can be seen from his diaries, edited by Else Wegener.[7] Often he stood between the incomprehensible stipulations of the authorities and our own needs which had been based on Wegener's own expedition plan. This is the chief source of friction in expeditions, since every member of the expedition wants to carry out as perfectly as possible the task he has been allotted, even when the overall situation calls rather for a more limited performance. We had all, thanks to Wegener's tireless help, got ourselves so inside the parts we were to play on the ice cap that we found it very hard to brook any economies which would make very little difference to the total cost of the expedition but which would considerably affect and lessen the final usefulness of our carefully planned work. The same argument was also raised by the English Greenland explorer R. A. Hamilton.[20] Wegener had the bitter lot of keeping the peace so that there was no breakdown before the journey commenced. Once out there, he knew, the logic of events would make its own imperative demands; or, as he said consolingly to me when a difference of opinion had arisen with the patron of the expedition about a point in the expedition's program: "We may sometimes have to give way despite our convictions, but when we come home triumphant, laden with new scientific knowledge, then such legal quibbles will be of as little importance as a scrap of paper!" I can remember no occasion during all these difficult situations when Wegener became angry and lost his self-control. We attributed this to his long years of collaboration with the deliberate Danish explorers; and he had learnt to use his diary instead as a kind of personal lightning conductor.

Thus he was able to write to me in a wonderfully human letter about the conflict mentioned above:

> You will certainly not, because of a passing ill mood, make a decision which for many years, perhaps for your whole life, would cloud the memory of our expedition. I believe rather—at least, I hope—that when you reach such a point you will bury the hatchet and grab a camera instead and reaffirm the principle which I have used to smooth away many a difficulty during my expeditions: Whatever happens, the cause must not suffer in any way. It is our sacred trust, it binds us together, it must go on under all circumstances, even with the greatest sacrifices. That is, if you like, my expedition religion, and it has been proved. It alone guarantees an expedition without recriminations.

Contrary to his expectations, in Greenland too there were more than enough causes for such entries in his diary, mainly because impenetrable winter ice blocked the coast delaying us by six weeks (see pages 35–46). Wegener himself was continuously skirting the ice margin in his motorboat *Krabbe* or standing in the crow's nest on the lookout for some place to slip through. He sat for days on end in the base camp, planning with the Greenlanders who were to help with dog sledges, the boat, or the ascent of the glacier. He negotiated with the leaders of the settlements at Umanq, Uvkusigsat, Ikerasaq, Nugaitsiaq, etc., over the catching and drying of shark for dogfood, the hiring of dog teams, the collecting of hay for our horses and about taking some strapping Greenlanders with us on our main journeys into the interior. He repeatedly made the long journey from the fjord to the edge of the ice cap at a height of almost 1,000 meters above sea level to test and improve the first road in Greenland constructed along a moraine and, most of all, to help and to encourage everywhere when unexpected hold-ups occurred. He checked time and again the flow of huge quantities of expedition material from one intermediate depot to the next and later on to the ice cap; he took part in the trial trips of the new propeller-driven sledges which he had been the first to bring into the Arctic—but the only thing that would have compensated him for the burden of his many duties as leader of the expedition would have been a journey with dog sledges to the "Mid-Ice" station in the best season of the year, and this, just because of his duties, he had to renounce. When, in late autumn, he finally thought that he would have to take additional provisions to this station himself by dog-sledge, enormous quantities of snow fell but, instead of a sudden storm, which would have packed the snow hard and made a wonderful sledge-track, the weather did not break. So this trip became an awful ordeal but also an achievement of which he and his two colleagues could be proud. Yet, amidst all this hectic activity, he retained the inner calm to talk to his fellow workers about their own personal problems at opportune moments, to discuss their future after their return home, and also to talk about not only scientific or technical matters, but also general literary, cultural, and human problems in the widest sense.

I shall not forget the last meeting I had with him when his successful journey eventually ended on 30th October 1930 at "Mid-Ice" with the temperature at −50 to −54°C. He greeted with beaming enthusiasm our dirty, low dug-out as a comfortable living room after the icy air in the tents during the preceding days, with joy shining in his face because Dr. Ernst Sorge and I, despite the lack of many provisions, were going to try to keep the station going until the following summer. He began enquiring at once about meteorological and glaciological observations and, with his thoughts already working on an enthusiastic report home, began to copy down in his notebook the most important temperature information. Now after he had, in his capacity as expedition leader, taken a look to see that things were all right there, he pushed back to his own glaciological problems in the western border region. On his fiftieth birthday, fully fit and with rested dogs, he set off, certain of making a quick march back in comparison with the tedious journey out. Even with this difficult journey ahead of him he did not forget our state of mind, our downheartedness because, due to the absence of necessary scientific equipment, our work must remain patchy. "The fact that you have spent the winter here in the middle of Greenland, even without any particular results in research, doing only the simplest and most routine measuring, is something which is worth all that has gone into the expedition," he said; and who knows if it was not this encouraging word that helped us psychologically to get through that winter?

Wegener himself would have opposed very strongly any attempts to make a hero out of him or to say that he never made mistakes. The best part of our memories of him is that in every respect he was close enough to us for true mutual understanding and yet at the same time he towered so far above us that we, in spite of his ever-friendly attitude, always looked up to him with respect.

Professor W. Wundt, a friend of his from student days, has summed up his memories of him thus:

> Alfred Wegener started out to tackle his scientific problems with only quite normal gifts in mathematics, physics, and the other natural sciences. He was never, throughout his life, in any way reluctant to admit this fact. He had, however, the ability to apply these gifts with great purpose and conscious aim. He had an extraordinary talent for observation and knowing what is at the same time both simple and important, and what can be expected to give a result. Added to this was a rigorous logic, which enabled him to assemble rightly everything relevant to his ideas. One can call this ambition, but it was a justified and legitimate ambition, which never led him to hurt or detract from anyone else. A noble mind and great loyalty were amongst his basic character traits. If he was somewhat authoritarian and obstinate in great matters, he shared this characteristic with all men who have achieved something significant in their lives. A compensating factor was his dry sense of humor that spared himself least of all.

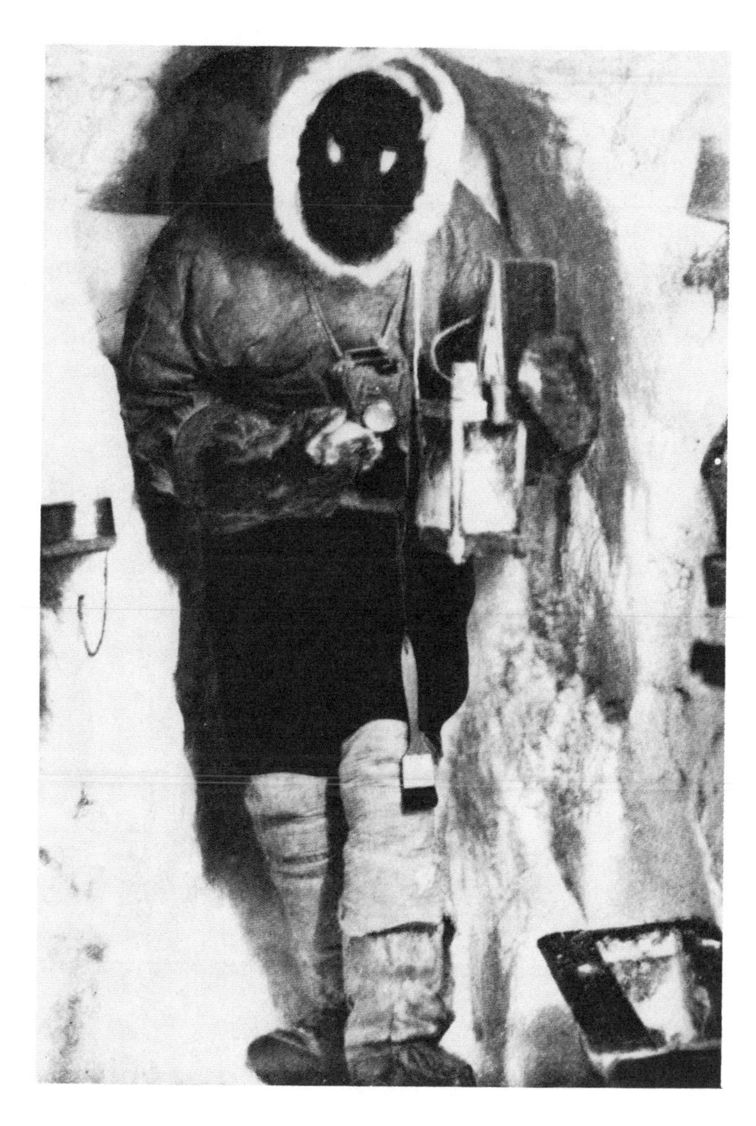

Johannes Georgi, prepared to take a weather observation during the 1930 expedition. He is wearing a mask to protect his face from driving snow. He carries two lamps and also a brush for removing snow from the instruments.

To this reminiscence of Wegener's early days let there be added as conclusion one from a friend of his later years, Professor Benndorf:[1] "I believe that Wegener took his military service (as an officer in World War I) very hard. Not because of dangers and hardships, because that would have appealed rather to a nature like his, but because of the difficult conflict it must have caused in him between his duty to his fatherland and his inmost conviction of the futility of war. Wegener was one of those rare men who do not willingly stop at duty for the good of self, family, or native land but who rather see in the promotion of the well-being of mankind as a whole the true purpose of life."

10

CONTINENTAL DRIFT AND PLATE TECTONICS: A REVOLUTION IN EARTH SCIENCE

BY I. BERNARD COHEN*

*Victor S. Thomas Professor of the History of Science, Harvard University

Continental Drift and Plate Tectonics: A Revolution in Earth Science

The recent revolution in earth science is notable for a number of features that illuminate the nature of all revolutions in science. But this revolution also exhibits some novel aspects that are peculiarly characteristic of the science of our times. Essentially, this revolution consists in the rejection of the traditional view that continents were formed or have grown or developed in fixed places, and the adoption of the radical notion of a "drift" of continents, with respect to one another and the ocean basins, on the surface of the earth. A feature of this revolution is the concept of plate tectonics: that the earth is divided into a set of rigid plates, containing continents and portions of the ocean floors, which rotate very slowly relative to each other.

The hypothesis of continental drift was put forth as a serious scientific proposal by Alfred Wegener (1880–1930) in 1912 and advanced in a major published monograph a few years later (Wegener 1915*). Recognized almost at once as potentially revolutionary, requiring a drastic revision of the very fundamentals of geologic science, the idea of continental mobility was widely discussed by geologists during the 1920s and 1930s

*These reference citations can be found at the end of this chapter beginning on page 196.

and was almost completely rejected. Wegener's proposal of continental drift thus produced only what I have called a revolution on paper until some time in the mid-1950s, when new kinds of evidence began to favor the possibility of a movement of continents. But it was not until the 1960s that a revolution in science actually occurred.

Historical analysis reveals that the revolution in science that ended the half-century status of revolution on paper was not merely a belated acceptance of a somewhat dormant or rejected set of ideas or theory. The revolution in science followed the introduction of new techniques for studying the earth and the broadening of knowledge by using new sources of information. Not only did many earth scientists begin to think along nontraditional lines, but there was an important input made by scientists trained in physics rather than geology. Hence the eventual revolution in earth science was not so much a mere revival of a long-rejected theory of continental drift as it was a radical transformation of the older idea, encompassing a newly conceived theory of plate tectonics to describe the motion. In a sense, Wegener's original theory never produced a revolution in science, but the eventual revolution in science did embody the central concept of continental mobility and the idea of two types of domains (continents and ocean floors) of Wegener's theory.

An outstanding feature of the recent revolution has been a general consciousness among working geologists that they have been living through a revolutionary moment in their science. A number of earth scientists have written articles or monographs stressing the revolutionary nature of the changes in thinking about continents and the earth, and have produced books with such titles as *A Revolution in the Earth Sciences: From Continental Drift to Plate Tectonics* (Hallam 1973) or *The Road to Jaramillo: Critical Years of the Revolution in Earth Science* (Glen 1982). This stress on revolution has not only been a feature of later historical or review articles and books but of working papers during the stages of revolution. For instance, a seminal technical paper in *Science* (Opdyke et al. 1966), entitled "Paleomagnetic Study of Antarctic Deep-Sea Cores," bore the subtitle "Revolutionary Method of Dating Events in Earth's History." In 1970, during a discussion of "a new class of faults," J. Tuzo Wilson remarked that recent discoveries concerning reversals of terrestrial magnetism constituted a "revolution" in earth science. In the final report (1972) of the Upper Mantle Project (U.M.P.) of the International Council of Scientific Unions, the "emergence of a unifying concept of plate tectonics" during the "period of the U.M.P." was said to be a "revolution" in the development of the earth sciences (Sullivan 1974, p. 343).

A consciousness of revolution in the historical reviews and summaries written during the early 1970s (primarily by English-speaking scientists) in some measure reflects the fact that the acceptance of continental motion and plate tectonics in the 1960s coincided with the considerable attention given to Thomas S. Kuhn's seminal tract, *The Structure of Scientific Revolutions*, published in 1962. Thus Allan Cox (1973), Anthony Hallam (1973), Ursula Marvin (1973), and J. Tuzo Wilson (1973, 1976) all refer specifically to Kuhn in reviewing or discussing the recent changes in theories about the continents. This revolution in science is also notable for having produced—in a little more than a decade after its occurrence—a series of well-documented histories or historical articles, many written by earth scientists, some of whom had themselves made fundamental contributions to the revolution.

As a result of these recent historical investigations, we now know that Francis Bacon was not the originator of the idea that continents have moved (Marvin 1973). Rather, he merely pointed out that there is a general conformance of the west coastal outlines of Africa and Peru. Nor did Alexander von Humboldt, almost two centuries later, advance from a recognition of the congruence of the facing coasts across the Atlantic to the suggestion that the two continents had once been joined and then moved apart. But the breaking-up of a protocontinent and the subsequent separation of its parts was suggested in 1859 in a marginally scientific book entitled *Creation and Its Mysteries Unveiled*, written in French by Antonio Snider-Pellegrini, an American living in Paris. Claims have also been advanced (wrongly, as pointed out by Ursula Marvin 1973, p. 58) for the Austrian geologist Eduard Suess as an early proponent of the idea of continental drift. But Suess did put forth, at the beginning of the twentieth century, the hypothesis that there had originally been two large Paleozoic continents: "Atlantis" (located in the North Atlantic Ocean) and "Gondwana-Land" (in the South). The latter he named for Gondwana (home of the Gonds), a district in central India. Suess—like some of his nineteenth-century predecessors—believed that our present continents were remnants of larger primitive ones, pieces of which had foundered into the ocean basins; but he did not propose that this process entailed a continental drift as we would understand that term today (Marvin 1973, p. 58).

A better case can be made for an American geologist, F. B. Taylor, who published in 1910 "a lengthy paper giving the first logically worked out and coherent hypothesis involving what we would now recognize as continental drift" (Hallam 1973, p. 3). First presented in a pamphlet in 1898, Taylor's theory was astronomical, not based on geography or geology. He hypothesized that a long time ago the earth had captured a comet which became our moon, an event which increased the earth's speed of rotation and generated a large tidal force; the combination of these two effects tended to pull the continents away from the poles. In his paper of 1910, and in later publications, Taylor elaborated his argument for continental movements by means of geological evidence, but these never attracted much attention in the geological community (Marvin 1973, pp. 63–64). In 1911 another American, Howard B. Baker, proposed that there had been a displacement of the continents, caused by astronomical forces, among them planetary perturbations in the solar system. When Wegener published his treatise, he summarized the work of a number of possible predecessors, including a lengthy paragraph discussing Taylor's contributions (1924, pp. 8–10). But Wegener twice asserted that he "became acquainted with all these works only when the displacement theory in its main outlines had already been worked out" (pp. 8, 10). In the last edition of his book, Wegener (1966, p. 3) added some new names to the historical register. He now said: "I also discovered ideas very similar to my own in a work of F. B. Taylor which appeared in 1910."[1]

Wegener's theory of continental motion

Serious discussion of the hypothesis of continental drift by geologists and geophysicists began with the publications of Alfred Wegener. He was by training and profession not a geologist but rather an astronomer and meteorologist (whose doctoral dissertation was in the domain of the history of astronomy).[2] Wegener's academic career embraced, first, a post in astronomy and meteorology at Marburg and, later, a professorship of meteorology plus geophysics at Graz (1924–1930). In his twenties and thirties Wegener went on meteorological expeditions to Greenland; he lost his life on a third such expedition in 1930. According to Lauge Koch, who went with Wegener on the first of these expeditions, the idea of a continental drift arose while Wegener was observing slabs of ice split apart in the sea. But Wegener himself says only that at about Christmastime in 1910 he was struck by the way the coastlines on the two sides of the Atlantic fit together and that this suggested to him the possibility of a lateral motion of the continents.

Apparently, Wegener did not take his own idea too seriously, dismissing it as "improbable" (Wegener 1924, 5; 1966, 1). But he began to develop the hypothesis of continental motion in the autumn of the following year, he relates, when "quite accidently" he came upon "a compendium of references describing the faunal similarities of Paleozoic strata in Africa and Brazil" (Marvin 1973, 66). In the compendium, the existence of identical or near-identical fossils of animals on both sides of the Atlantic in the distant past was used in the then-traditional manner to argue for the existence of ancient land bridges between Africa and Brazil. Snails, for example, simply could not have swum across the vastness of the Atlantic Ocean; hence to find the same or very similar fossil snails on both sides of the South Atlantic carries a strong probability that there must have been some connection by land between South America and Africa long ago. The alternative would be to assume a similar but independent evolutionary development on a large scale in the two regions—a highly improbable occurrence.

Wegener was much impressed by the fossil evidence, but he rejected the hypothesis that the continents had once been connected by some kind of land bridge or now-sunken continent; this hypothesis required the further supposition of the sinking or disintegration of these land connections, for which there was no scientific evidence. Of course there are land bridges between continents, for example the isthmus of Panama and a former bridge across the Bering Strait, but there was no real evidence for a hypothesized land bridge spanning the South Atlantic in ancient times. As an alternative theory, Wegener revived his earlier ideas of the possibility of continental drift, transforming them from what, as he confessed, was only a "fantastic and impractical" idea—no more than a jigsaw puzzle solution of no real significance for earth science—into a working scientific concept. Wegener developed his hypothesis, adducing various kinds of supporting evidence, and summarized his results in 1912 at a meeting of geologists. His first two reports were published later in that year, and in 1915 he published a monograph, *Die Entstehung der Kontinente und Ozeane (The Origin of Continents and Oceans)*, in which he marshaled all the evidence he had found to support his ideas. This book appeared in revised versions in 1920, 1922, and 1929 and was translated into English, French, Spanish, and Russian. In the English version (1924), based on the third German edition of 1922, Wegener's expression "Die Verschiebung der Kontinente" was accurately rendered as 'continental displacement', but almost at once the phrase generally used was 'continental drift.'[3]

Wegener based his argument on geological and paleontological evidence, not mere pattern-fitting. He put great stress on aspects of geologic similarity on both sides of the ocean. In the last edition of his book, he adduced supporting evidence from the field of paleoclimatology, a subject on which he had written a book in 1924 (in collaboration with W. Köppen), inferring a wandering of the poles of the earth. Wegener postulated that in the Mesozoic era, and continuing up to the fairly recent past, there had been a huge supercontinent or protocontinent, which he named Pangaea, that had broken apart. The fragments of Pangaea separated and moved away from one another, producing the individual continents we know today. Two possible causes of a drift or motion or displacement of continents were put forth: tidal forces produced by the moon, and a force of "flight from the poles" (*Pohlflucht*), that is, a kind of centrifugal effect produced by the rotation of the earth. But Wegener was aware that the solution to the problem of the cause of continental movement still eluded him. "The Newton of drift theory," he wrote, "has not yet appeared," thus echoing the sentiments of Cuvier, van't Hoff, and others. "It is probable," he admitted, "that the complete solution of the problem of the driving forces will still be a long time coming." In retrospect, Wegener's most fundamental and original contribution was his concept that the continents and ocean floors were two distinct layers of the earth's crust, differing from each other both in rock composition and in altitude. Most scientists at that time believed that the oceans, with the exception of the Pacific, had sialic floors. Wegener's basic ideas were validated by plate tectonics.

Though Wegener's theory of continental drift was only a revolution on paper for some time, this does not mean that his ideas attracted no attention and no followers. Far from it! The decade of the 1920s was marked by a series of violent international controversies. An unsigned review of the second edition of Wegener's book (1920), in the influential journal *Nature* on 16 February 1922 (vol. 109, p. 202), gave an adequate summary of the main points of the theory and expressed the hope that an English version might soon appear. Noting that there was "strong opposition from a good many geologists," the reviewer concluded that if Wegener's theory could be substantiated, a "revolution in thought" would occur similar to "the change in astronomical ideas at the time of Copernicus" (p. 203). A generally favorable report on a lecture by Wegener appeared in 1921 in the most important German scientific journal, *Die Naturwissenschaften* (1921, 219–220). The reporter, O. Baschin, said that those who heard the lecture, at the Geographical Society of Berlin, were "extraordinarily convinced" and that the theory aroused "general approval," although there were some objections and warnings of caution in the ensuing discussion. Baschin concluded that there "is no factual proof against Wegener's theory," but that "solid proof must still be found before it can be unreservedly accepted."

Quite different in tone was the review in the British *Geological Magazine* for August 1922, in which Philip Lake stated bluntly that Wegener "is not seeking truth; he is advocating a cause and is blind to every fact and argument that tells against it." In America, in the October 1922 issue of *Geographical Review*, Harry Fielding Reid delivered what he saw as the coup de grâce to both continental drift and polar wandering. In the autumn of that same year, the continental drift hypothesis was the subject of debate and discussion at the annual meeting of the British Association for the Advancement of Science; the published report (by W. B. Wright) described the event as "lively but inconclusive." But a favorable presentation of "The Displacement of Continents: A New Theory" was published by Professor F. E. Weiss in the *Manchester Guardian* on Thursday, 16 March 1922. Weiss said Wegener's theory was "of fundamental importance to the sciences of geography and geology" and "of great interest to the biological sciences," and he concluded that the theory "constitutes a good working hypothesis" which will "greatly stimulate further inquiry."

A major scientific event of the 1920s was the debate in New York in 1926, convened by the American Association of Petroleum Geologists and published under the title *Theory of Continental Drift: A Symposium on the Origin and Movement of Land masses. . .as Proposed by Alfred Wegener* (van der Gracht 1928). Among those in attendance were Frank B. Taylor and eleven other participants including eight Americans and three Europeans. The chairman, W. A. J. M. van Waterschoot van der Gracht (a Dutch geologist, vice president of Marland Oil Co.), contributed to the published proceedings a lengthy introduction supporting continental drift and a concluding rebuttal to the opponents; these two papers occupy more than half of the volume. Some of the participants (Chester Longwell of Yale, John Joly of Dublin, G. A. F. Molengraaf of Delft, J. W. Gregory of Glasgow, Joseph T. Singewald, Jr. of Johns Hopkins) were somewhat tolerant but highly skeptical, while others (Bailey Willis of Stanford, Rollin T. Chamberlin of Chicago, William Bowie of the U.S. Coast Guard and Geodetic Survey, Edward

W. Berry of Johns Hopkins) larded their geological counter-evidence with sarcastic remarks about the faulty science and defective method which they claimed characterized Wegener's thinking and writing. Looking back from today's vantage point, what is most impressive is the rancor and virulence in these criticisms.[4] It is obvious that Wegener had launched what seemed to be a frontal attack on the foundations of earth science and of sound belief.

The Wegener hypothesis aroused hostility on a number of grounds. First, it went directly counter to the mind-set of almost all geologists and geophysicists, who had been conditioned from their earliest days to think of the continents as essentially stable, of the earth as terra firma. To suggest that continents might have a lateral motion with respect to one another was as "heretical" and "absurd" in a laic sense as the Copernican hypothesis had been in the time of Galileo.[5] Second, the new hypothesis supposed that the earth is not as rigid as had been believed and as seemed evident even to the most superficial observer. Hence, there seemed to be required some imagined forces of tremendous magnitude, by far exceeding—as geophysicists like Harold Jeffreys were quick to point out—the two proposed by Wegener. The argument could be boiled down to the impossibility of having what has been called "a weak continental ship" sail "through an unyielding oceanic crust" (see Glen 1982, 5).

It is unfortunately common to attempt to dismiss a proposed revolution in science by making *ad hominem* comments about the proponent of the new theory. Wegener not only was attacked for his method but was denied the right to discuss geology because he lacked credentials, being a meteorologist rather than a geologist—and a German meteorologist at that. Charles Schuchert, emeritus professor of paleontology at Yale (Van der Gracht 1928, 140), referred to continental drift as a "German theory," and he quoted with evident approval the remarks of Pierre Termier (director of the Geological Survey of France) that Wegener's theory was only "a beautiful dream," the "dream of a great poet"; when one "tries to embrace it," one "finds that he has in his arms but little vapor or smoke." Wegener, furthermore, according to Schuchert, "generalizes too easily" and "pays little or no attention to historical geology" (p. 139). He was an outsider, who had never done any actual field work in paleontology or geology; it is "wrong," Schuchert concluded, "for a stranger to the facts he handles to generalize from them to other generalizations."

That Wegener was rejected—at least in part—because he was not a member of "the club" is borne out by the literature. In his attack on Wegener's theory, evidence, and method of scientific procedure, Harold Jeffreys in *The Earth: Its Origin*, (Cambridge, 1952, p. 345) declared that "Wegener was primarily a meteorologist." In 1944, in an article in the *American Journal of Science* (vol. 242, p. 229), Chester R. Longwell wrote condescendingly that "charitable commentators" suggest that Wegener's inconsistencies and omissions "may be overlooked on the ground that [he]. . .was not a geologist." And, as late as 1978, George Gaylord Simpson (1978, 272) repeated his earlier opinion that "most of Wegener's supposed paleontological and biological evidence" was either "equivocal" or "simply wrong"; he criticized Wegener (a "German meteorologist") for daring to venture into fields with which he "had no firsthand acquaintance."

Most geologists in the 1930s and 40s would have agreed with Jeffreys when he said in the third edition of his influential treatise, *The Earth* (1952, 348), that "the advocates of continental drift have not produced in thirty years an explanation that will bear inspection." Orthodox geologists and paleontologists even used the idea of continental drift to "supply comic relief" in their classroom lectures. Percy E. Raymond, professor of paleontology at Harvard, told his students how "half of a Devonian pelecypod" had been found in Newfoundland and another half in Ireland. The two "matched perfectly" and so must have been "the two halves of the same pelecypod, which had been wrenched apart by Wegener's hypothesis in the late Pleistocene" (Marvin 1973, 106).

The twenties and thirties were not wholly devoid of Wegener supporters, however. Reginald A. Daly of Harvard favored the general idea of continental mobility, although he was far from being a strict Wegenerian, having once himself referred to Wegener as "a German meteorologist" (1925). Daly produced his own version of the motion of continents, which, in retrospect, began "an approach to the current model of plate tectonics" (Marvin 1973, 99). On the title page of his book, aptly called *Our Mobile Earth* (1926), Daly placed the words "E pur si muove!" a version of the statement allegedly made by Galileo after recanting his Copernican allegiance to the theory of a moving earth ("Eppur si muove"; "Nevertheless it does move").[6]

One of the primary supporters of Wegener's ideas during the 1920s was Emile Argand, founder and director of the Geological Institute at Neuchâtel, Switzerland. In 1922, at the first post-World War I international congress of geologists, Argand delivered a spirited defense of Wegener's fundamental concept in relation to "Tectonics of Asia." Argand not only gathered and organized an impressive mass of evidence supporting Wegener, but also performed a valuable function in making a distinction between a Wegenerian kind of "mobilisme" and orthodox "fixisme." He declared that "fixisme" "is not a theory but a negative element common to several theories" (Argand-Carozzi 1977, 125). Although he made an affirmative exhibition of the variety of details of the arguments favoring "mobilisme," Argand could not help admitting in his conclusion that "almost nothing is known about the forces responsible for continental drift" (p. 162).

Wegener's two chief supporters during the 1930s were Arthur Holmes, regarded by many "as the greatest British geologist of this century" (Hallam 1973, 26), and the South African geologist Alexander du Toit. Holmes was a convert by the time of the appearance of the American symposium on continental drift, a volume he reviewed in *Nature* (September 1928). Here he noted that "all the adverse criticism is. . .mainly directed against Wegener" rather than against Wegener's ideas in general, that "when all has been said, there remains a far stronger case for continental drift than either Taylor or Wegener has yet put forward." Although he accepted the general idea of the motion of continents and became Britain's major advocate of continental drift, Holmes proposed a new mechanism for producing this effect, according to which convection currents in the earth's mantle (the part of the earth immediately below the crust) would tend to produce mountain-building and continental drift (see Marvin 1973, 103; Hallam 1973, 26ff.). Du Toit lived in Johannesburg, "in the heart of ancient Gondwana," as Ursula Marvin reminds us (1973, 107), "where the evidence of continental drift is most compelling." He summed up his ideas in a work called *Our Wandering Continents* (1937), subtitled "an hypothesis of continental drifting." It was dedicated "To the memory of Alfred Wegener for his distinguished services in connection with the geological interpretation of Our Earth." In this book, du Toit proposed a theory of the earth that differed somewhat from Wegener's (see Marvin 1973, 107–110; Hallam 1973, 30–36). For instance, instead of Wegener's original single continent, Pangaea, du Toit proposed two—Laurasia in the northern hemisphere and Gondwanaland in the southern hemisphere.

Du Toit attributed the rejection of Wegener's hypothesis to two factors: (1) the lack of a satisfactory mechanism for producing the drift, and (2) the "deep conservatism" which he found to be characteristic of the whole history of geology.

Yet du Toit was fully aware that the acceptance of continental drift would "involve the rewriting of our numerous text-books, not only of Geology, but Palaeogeography, Palaeoclimatology and Geophysics" (p. 5). He had no doubt, he said, that "a great and fundamental truth" was embodied in the "postulated Drift" and that Taylor and Wegener had proposed a "revolutionary Hypothesis" (p. vii).

Du Toit was not the only person to view Wegener's work as revolutionary. The reporter in *Die Naturwissenschaften* in 1921, the anonymous reviewer in *Nature* in 1922, F. E. Weiss in 1922, van der Gracht and others in 1926—friends and foes alike—used the term. Daly (1926, 260) characterized continental drift as a "new, startling explanation" which many "geologists have found. . .bizarre" and even "shocking," a "revolutionary conception." The same revolutionary character of "ideas so novel as those of Wegener" was manifest in Philip Lake's pronouncement that a "moving continent is as strange to us as a moving earth was to our ancestors" (1922, 338). In reviewing the proceedings of the New York symposium of 1926 (van der Gracht et al. 1928) in *Geological Magazine* (1928, vol. 65, pp. 422–424), Lake explicitly referred to Wegener's "revolutionary theory."

Wegener himself was fully aware of the revolutionary potential of his new ideas. Writing in 1911 to W. Köppen, a year before he made any public statement of his new ideas in lectures or in print, Wegener asked why we "should hesitate to throw the old view overboard?" He continued: "Why should one hold this idea back for 10 or even 30 years? Is it perhaps revolutionary? [Ist sie etwa revolutionär?]" He appended a bold and simple answer to his rhetorical question: "I do not believe the old ideas have more than ten years to live."[7]

Because of its revolutionary qualities, stronger evidence than usual would have been necessary in order for this theory to win support from the scientific community. For any funda-

zeigt, daß jetzt Sinn und Verstand in die ganze geologische Entwicklungsgeschichte der Erde kommt, warum sollen wir zögern, die alte Anschauung über Bord zu werfen? Warum soll man 10 oder gar 30 Jahre mit dieser Idee zurückhalten?[x] Ist sie etwa revolutionär?

x) Ich glaube nicht, daß die alten Vorstellungen noch 10 Jahre zu leben haben. Noch ist die Lehre von der Isostasie nicht vollständig durchgearbeitet. Sobald sie es sein wird, wird der Widerspruch zu Tage liegen und die alten Vorstellungen werden korrigiert werden

Excerpt from a letter from Wegener to Köppen written in 1911. "[Wenn sich] zeigt, daß jetzt Sinn und Verstand in die ganze Entwicklungsgeschichte der Erde kommt, warum sollen wir zögern, die alte Anschauung über Bord zu werfen? Warum soll man 10 oder gar 30 Jahre mit dieser Idee zurückhalten? Ist sie etwa revolutionär? Ich glaube nicht, daß die alten Vorstellungen noch 10 Jahre zu leben haben. . . ."

mental and radical change to be accepted by scientists, there must either be unassailable or compelling evidence, or there must be a whole set of obvious advantages over existing theories. It seems clear that in the 1920s and 30s, neither of these conditions had as yet obtained. In fact, such unassailable evidence did not become available until the 1950s and after. And in the meantime the radical restructuring of the whole science of geology which would be required for the acceptance of Wegener's ideas did not seem to be at all attractive in the absence of more compelling facts. The Chicago geologist R. T.

Chamberlin reported at the 1926 symposium of the American Association of Petroleum Geologists that at an earlier meeting of the Geological Society of America (held in Ann Arbor in 1922) he had heard it said that "if we are to believe Wegener's hypothesis, we must forget everything which has been learned in the last 70 years and start all over again." In retrospect, this has proved to be perfectly true. Chamberlin's remarks, we may observe, were restated in a somewhat different context some forty years later when J. Tuzo Wilson (1968a, 22) wrote, "If indeed the Earth is, in its own slow way, a very dynamic body, and we have regarded it as essentially static, we need to discard most of our old theories and books and start again with a new viewpoint and a new science."

As for Wegener's failure to provide a satisfactory mechanism for continental drift—generally agreed to be "the greatest stumbling-block to acceptance of Wegener's hypothesis"—Hallam (1973, 110) tries to counter such an argument by pointing out that "gravity, geomagnetism, and electricity were all fully accepted long before they were adequately 'explained.'" He adds that in geology the lack of any "general agreement" about the "underlying cause" has not prevented the universal acceptance of the "existence of former ice ages." And J. Tuzo Wilson (1964, 4) argued, has not man always been willing to accept the existence of phenomena (for example, the earth's magnetic field) or of processes (such as thunderstorms) "long before he could account for them?" At this point, however, a clarification is required. Rachel Laudan has wisely pointed out that the problem of *cause of 'mechanism'* with respect to continental drift differs greatly from the situation with respect to "gravity, geomagnetism, and electricity," or "the existence of thunderstorms." In the case of continental drift, the problem "was not simply that there was no *known* mechanism or cause," but rather that "any *conceivable* mechanism would conflict with physical theory" (R. Laudan 1978; see Gutting 1980, 288).

Furthermore, there existed unifying theories of the nature of the earth and its interior which explained most observations tolerably well.

S. K. Runcorn (Gutting 1980, 193) observes that whereas in the 1950s and earlier "the absence of a mechanism" was "widely held to be a most weighty objection to accepting the geological or palaeomagnetic evidence for continental drift," today "the "theory" of plate tectonics is [becoming] accepted without any consensus about the physical mechanism responsible." He sees, as of 1980, "the question of the nature of the forces that move the plates" as "the most important challenge facing geophysicists today."

Transformations of Wegener's theory

In retrospect, now that we have witnessed the revolution, there appear to have been two radical innovations that separate the age of the revolution on paper from the present. The first is the accumulation of new and convincing evidence that continents and ocean floors are indeed real entities that have moved with respect to one another—evidence that by several orders of magnitude transcends the fitting of coastlines and even the matching of transoceanic geological or ecological features and of fossil flora and fauna. The second is a radical restatement of the theory that has so altered the fundamental concepts as to make it problematic whether the revolution that was ultimately achieved can legitimately be identified as the attempted revolution which failed for almost half a century. This situation has many points of similarity with that of the so-called Copernican revolution. The eventual revolution, set into motion by Galileo and Kepler and achieved by Newton, kept only the most general Copernican cosmological idea, that the earth moves and the sun stands still, while rejecting the essential features of Copernicus's astronomy. Similarly, the evolution in earth science kept only Wegener's most general idea, that there may be a motion of continents with respect to one another, while rejecting the essential features of Wegener's theory—that continents (made of sial) move on or through the oceanic crust (sima) as individual or separate entities, while the denser envelope of sima remains fixed in location.

The current idea is rather that large blocks or plates covering the earth's surface move and that some of these may carry continents or parts of continents, together with ocean floor. Hence a theory of individual continents in motion has been replaced by a different theory in which the motion of the continents is but a visible part of a more fundamental motion. In the process, Wegener's hypothesizing of a "Pohlfluchtkraft" and a tidal force became irrelevant.

The new evidence from the 1950s came, in the first instance, from research on paleomagnetism and on the magnetism of the earth. Paleomagnetism is the study of "remnant" rock magnetism, the magnetism that remains in a rock sample as it solidifies from a molten state. This magnetism is imprinted into iron oxides in the rock as an effect of the earth's magnetic field. Studies made by P. M. S. Blackett in London and by S. K. Runcorn in Newcastle (later in Cambridge) and others showed that the magnetic field of the earth has not been constant but has changed—even undergoing reversals—according to a time pattern that can be determined. (These studies were made possible by a sensitive new instrument, the magnetometer, of which Blackett was the primary inventor). When the path of the position of the magnetic poles had been carefully plotted, it was found that the movement of the pole differed from one place to another, suggesting that each of these land masses had moved independently. The evidence pointed toward a time when the southern continents had been joined together in the south polar region as a protocontinent, Gondwanaland, so that there must have been lateral displacements of its parts—our present continents (see McKenzie 1977, 114–117).

These first results did not at once convince the community of earth scientists that there had been a continental displacement; no doubt, there were too many unsolved problems about details of the history of the earth's magnetic field, while the argument from magnetism was of a "complicated and specialist nature" with many untested assumptions (MacKenzie 1977, 116). But sufficient interest was aroused, chiefly among geophysicists, that in 1956 a symposium was held on the subject of continental drift. E. Irving of the Australian National University reviewed the magnetic studies of the past years and concluded that "the balance of the evidence favors the idea that. . .the Earth's axis [has] changed its position relative to the Earth as a whole [and] that also the continents have 'drifted' with respect to each other" (Irving 1953; 1958; see Marvin 1973, 150–152).

A second line of research that tended to advance the revival of Wegener's basic idea, although not Wegener's theory, was the study of mid-ocean ridges. The oceans and inland seas cover about 70 percent of the surface of the earth; therefore, since knowledge of the nature of the regions below the oceans was rudimentary in the 1930s and 40s, we can understand why the pre-war debate concerning continental drift had turned out to be so inconclusive (Hallam 1973, 37). Yet the mid-Atlantic ridge had been mapped and in 1916 F. B. Taylor said it appeared as though the Atlantic continents had crept away from the ridge to either side. Wegener, furthermore, did write out the evidence that the ocean floors were basaltic—arguing from density, magnetism, and compositions, and so on—but nobody paid attention. The direct key to our current knowledge of the motion of continents came from studies of the earth beneath the oceans. During the International Geophysical Year (1957–58), new techniques were introduced for measuring gravity and for combining seismic and gravity data. Geophysicists found ways to determine the rate at which heat flows through the floor of the ocean. The implication of these studies was that huge blocks of oceanic crust could "apparently be displaced large distances with respect to each other" (Hallam 1973, 52). The congruence of these findings with those obtained from magnetic studies gave strong support to the notion that continents had undergone displacement with respect to one another. At this time, a model of continental drift was put forth in the now widely accepted concept of plate tectonics,[8] a system of architecture according to which the earth's crust is analogous to a "mosaic of large plates—akin, on a mega-scale, to ice floes or paving stones." These plates move as separate rigid units, and experience deformation at their borders with other plates. Ursula Marvin especially stresses the fact that these "moving plates are not the continents, as postulated by Alfred Wegener, nor are they the individual ocean floors" (1973, 165). Since each of these plates may include both continents and ocean floors, their motion is very different from Wegener's conception of moving continents. Hence, the original term 'continental drift', implying a motion of individual continents, is no longer strictly accurate (Hallam 1973, 74). In 1968, it was proposed that six large plates and twelve subplates cover the earth. Since then, additional subdivisions have been suggested.

Sea-floor spreading

The theory of plate tectonics is bound up with the concept of 'sea-floor spreading' to explain the instability and changes in the earth's crust. Proposed originally in 1960 by Harry H. Hess of Princeton, sea-floor spreading postulates a process whereby the ocean floor is constantly being pushed out on both sides of the ridge that runs through the major oceans.[9] Hess first circulated his ideas in 1960 in the draft of a chapter for a book on oceans which was not published until 1962. His central concept was so radical that he introduced it by saying that he considered his chapter to be "an essay in geopoetry." He proposed that the great ridges running down the center of oceans are outlets for the up-pouring of molten material from the earth's mantle, the region below the crust. This matter spreads out equally on both sides of the ridge, where it cools and solidifies, so as to become part of the old crust to which it becomes attached. As the crust grows in this way alongside the ridge, the material (a giant plate) moves laterally away from the ridge. Since the earth does not expand, it follows that this plate also cannot simply expand with the addition of new matter. Hence there must be a region away from the ridge where the plate disintegrates. That is, at the boundary furthest from the ridge, the plate slides beneath another plate, descending down into the mantle in an oceanic trench where it releases its water and becomes molten once again. This process has been linked to a kind of convection 'conveyor belt', bringing up matter from the mantle at a ridge, forcing it forward or away from the edge, and eventually taking the matter of the mantle back into the mantle itself at a distant trench.

There is thus a continual enormous pressure that pushes away from the mid-Atlantic ridge the two plates that carry Africa and South America. About 180 million years ago, South America and Africa were joined, forming the continent of Gondwanaland. The line of cleavage between the continents coincides with the still-active ridge that causes the spreading. This line is marked by the occurrence of earthquakes and is today roughly equidistant from the Atlantic boundaries of South America and Africa.

Hess thus provided what is generally considered to be a major element in the subsequent development of plate tectonics and our understanding of continental drift by showing the manner in which the ocean floor is constantly being created on one side of a continent and destroyed on the other (McKenzie 1977, 117). The speed with which the floor of the

Atlantic moves out from the ridge is about four centimeters (an inch and a half) per year, so that the time required for the ocean floor to well up at the midocean ridge, travel across the ocean, and finally descend back into the mantle is about two hundred million years. This number at once explained a number of mysteries. For instance, fossils from borings in the ocean floor were never older than some two hundred million years (the age of the Mesozoic era), even though marine fossils dug up on land showed that marine life dated back to at least ten times that age. Again, if the ocean beds were as old as the continents then the normal rate of sedimentation would have had to produce deep layers of sediment, but dredging expeditions had found that there is little sediment occurring on the ocean floors. In short, during the several billion years of existence of the oceans (Uyeda 1978, 63), the ocean floors have not been constant but continually changing, continually moving.

When Hess's ideas are combined with the general notion of plate tectonics, another process may be envisioned whereby the addition of new matter at the boundary of a plate does not increase its size. The plate can be continually shortened by compression, which manifests itself in the formation or alteration of mountain ranges where two plates collide.

In presenting his ideas on sea-floor spreading, Harry Hess was explicit on the fact that his theory was "not exactly the same as continental drift" (1962, 617). According to all continental-drift thinking, "continents. . .flow through oceanic crust impelled by unknown forces," but he put forth the idea that the continents "rather. . .ride passively on mantle material as it comes to the surface at the crest of the ridge and then move laterally away from it."

I have mentioned the general agreement that the initial purely paleomagnetic evidence did not fully convince most earth scientists of the need to abandon the concept of 'fixism'.

The final evidence came from new magnetic studies that dramatically confirmed the hypothesis of sea-floor spreading. Shipborn magnetometers revealed that there are areas of the ocean floor which are magnetized in strips (see Hurley 1959, 61). If Hess's explanation was correct, then there ought to exist a symmetry with respect to the magnetic strips on both sides of an oceanic ridge. This was the test proposed by F. J. Vine, then a research student at Cambridge University, and his supervisor, D. H. Matthews. Measurements soon proved that there was just that predicted symmetry.

According to theory, as hot molten material spreads out on both sides of an oceanic ridge and solidifies, it acquires the current magnetism of the earth. Though it is pushed out from the ridge by new matter, it retains the magnetism it acquired while cooling. Thus each successive stripe of matter should carry a magnetic imprint of the date of its formation into a solid mass, and this should be the same in both directions, on both sides of the ridge. This important hypothesis was published by Vine and Matthews in *Nature* in 1963 and successfully passed the test of experiment in the following years. In fact, the history of the earth's magnetism shows not only minor changes but definite reversals at now-known dates, all of which were actually found in successive stripes.

Though this hypothesis may sound very logical and unexciting today, it was in fact radical and daring at the time. Vine recalls that when he first mentioned his idea to Maurice Hill, a Cambridge geophysicist, Hill "thought I was totally mad," according to Vine, although he "was polite enough not to say anything; he just looked at me and went on to talk about something else" (Glen 1982, 279). Vine also talked about his hypothesis with Sir Edward Bullard, who "was much more encouraging," even though "he realized it was a bit of a long shot." Vine was "quite keen to publish the idea with Teddy Bullard," thinking "it would look great: 'Bullard and Vine.'

Teddy Bullard quite rightly said 'no way.' He didn't want his name on the paper." Bullard was a major innovator in geophysics who made important contributions to the theory of heat flow in the earth. Always receptive to new ideas, he "received the hypothesis with enthusiasm and spoke favorably of it to many," despite his unwillingness to accept Vine's initial offer of co-authorship (p. 358).

The hypothesis proposed by Vine and Matthews (and, independently, by Lawrence Morley in Canada; for details see Glen 1982, 271 ff.) was "equal in importance to any formulated in the geological sciences in this century" (p. 271). In addition to independently confirming Hess's idea of sea-floor spreading, it could enable the speed of spreading to be calculated on the basis of an independent, accurately determined time scale for the reversals of the earth's magnetic field. There seems to be agreement that the confirmation of the Vine-Matthews-Morley hypothesis "triggered" the "revolution" in the earth sciences (p. 269). The next step was the "formulation of a new theory of global tectonics" (Hallam 1973, 67) and the reconstruction of our knowledge of the earth.

Acceptance of the revolution

Whoever examines the history of the past twenty years of earth science, as presented in such works as Sullivan (1974) and especially Glen (1982), will recognize how much significant research was required in order to complete the revolution (including the heroic contributions, which I have scarcely mentioned, of such scientists as Edward Bullard, J. Tuzo Wilson, Maurice Ewing, and others).[10] Furthermore, many eminent geophysicists long refused to accept not only the theory of plate tectonics but even the concept of sea-floor spreading. "Not a single aspect of the ocean-floor spreading hypothesis can stand up to criticism," wrote Vladimir V. Beloussov in 1970, (Sullivan 1974, 105) a man described in *Nature* as "the Soviet Union's most distinguished geophysicist." A dozen years later, in December 1982, in the *Geophysical Journal of the Royal Astronomical Society* (vol. 71, 555–556), Sir Harold Jeffreys, aged ninety-one, for six decades the leading arch-foe of continental drift, still scorned the theory and likened the "subduction of oceanic plates" to "cutting butter with a knife also made of butter."

The conservatism of earth scientists, to which du Toit referred, is further illustrated by the article on Wegener in the authoritative *Dictionary of Scientific Biography* (14: 214–217), published in 1976. Written by K. E. Bullen of the University of Sydney, this account concludes with a grudging presentation of the evidence (from paleomagnetism and studies of the ocean floor) that has tended to support the rebirth of Wegener's ideas, followed by a listing of the arguments—old and new—"against continental drift": while "ad hoc answers to all these questions have been put forward by protagonists of the theory, yet again their answers have been questioned" (p. 216). In 1976, three years after the publication of Hallam's and Marvin's histories (both of which proclaimed the success of the revolution and analyzed its structure), the final judgment in the current biography of Wegener is that the "enthusiasms of a considerable number of earth scientists lead them to assert, sometimes with a religious fervor, that continental drift is now established" (p. 217).

Bullen's reference to "religious fervor" is especially apt because the language of the changes in opinion during the 1960s carried strong overtones of religious metaphor, especially in relation to conversion. This is a rather general feature of scientific revolutions, as T. S. Kuhn has observed. A typical example is the experience of J. Tuzo Wilson who, in 1959, was a primary antagonist of the theory of continental drift. But within a few years, Wilson had undergone conversion and referred to himself as "a reformed anti-drifter" (see Wilson 1966, 3–9, for a discussion of his subsequent conversion). He not only then developed some of the fundamental geological evidence favoring continental drift but became the major herald of the revolution. In 1963, at a Symposium on the Upper Mantle Project in Berkeley during the XIIIth General Assembly of the International Union of Geology and Geophysics, Wilson boldly announced that "earth science is ripe for a major scientific revolution" (Takeuchi 1970, 244). The present situation in earth science, he said, was "like that of astronomers before the ideas of Copernicus and Galileo were accepted, like that of chemistry before the ideas of atoms and molecules were introduced, like that of biology before evolution, like that of physics before quantum mechanics."

In the late 1960s and early 1970s, there were a large number of symposia on continental drift or on allied topics such as paleomagnetism and geomagnetism. One of these, part of the Annual General Meeting of the American Philosophical Society in April 1968, was entitled "Gondwanaland Revisited: New Evidence for Continental Drift." Here, in a paper entitled "Static or Mobile Earth: The Current Scientific Revolution," Wilson suggested (1968, 309, 317) that this "major scientific revolution in our own time" should "be called the Wegenerian revolution" in honor of its "chief advocate." Fellow earth scientists have generally agreed with Wilson that there has been a revolution and that Wegener should be honored as a first major articulator of the concept of continental drift. But this eponymous honor has not as yet been given to Wegener as it has to Copernicus, Galileo, Newton, Lavoisier, Darwin, and Einstein.

As we have seen, many writers have compared the revolution initiated by Wegener to the Copernican revolution. And it is similar, in that the eventual revolution in earth science was as far removed from the original theory of Wegener as the system of Kepler, Galileo, and Newton was from what was actually proposed by Copernicus. Just as there was no revolution in astronomy until more than a half-century after the publication of Copernicus's book in 1543, so there was no revolution in geological science until about a half-century after the publication of Wegener's original paper and his book. The eventual revolution that has been called Copernican was really a Newtonian revolution, based upon the achievements of Galileo and Kepler, and the 'Copernican system' that was at the foundation of the revolution was the system of Kepler. In like fashion, the eventual revolution that occurred in the earth sciences in the 1960s does not embody Wegener's theory but only his basic idea that the continents have not been fixed in their present place throughout all the earth's history and that they once clustered around the poles. Wegener's main contribution, the idea of mobilism as opposed to stabilism, played a role much like that of Copernicus's main contribution, the vision that it is possible to build a system of the world upon the concept of a moving rather than a stationary earth.

The general shift in earth science from stabilism to mobilism—specifically to ideas of continental drift and plate tectonics—is undoubtedly a revolution for the following reasons: First, this change in accepted geological opinion has been recognized as a revolution by major observers *at the time*, including practitioners in this subject area. I take this to be a primary index of the occurrence of a revolution in science. Second, an examination of the contents of the science before 1912 and 1970 shows the magnitude of the change to be sufficient for a revolution. Third, critical historians have concluded that the changes in the framework of geological thinking are of sufficient magnitude to constitute a revolution. Admittedly, this is a somewhat subjective judgment, but it serves as a corroboration of the conclusions of geologists and geophysicists participating in the revolution. We have seen that long before there was a successful revolution (that is, before there was a revolution in science, as opposed to a revolution on paper), a number of geologists—even those who opposed Wegener's ideas—were aware of the revolutionary character of the concept of continental drift and understood the revolutionary implications that its acceptance would have for the whole of geological science. Fourth, earth scientists today generally agree that a revolution has occurred in their discipline.

But what is the magnitude of the revolution? Is it a major revolution, to be compared to the Darwinian revolution, the revolution of quantum mechanics and relativity, or the Newtonian revolution? Or is it a revolution on a smaller scale, more akin to the Chemical Revolution? We have seen that George Gaylord Simpson called it a "major sub-revolution." D. P. McKenzie (1977, 120–121), in the conclusion of an article, "On the Relation of Plate Tectonics to the 'Evolution' of Ideas in Geology," contrasted the "impact of plate tectonics on the geological sciences" with the impact of "the discovery of the structure of DNA in the biological sciences." He concluded that "plate tectonics was less fundamental a revolution than the discoveries which began molecular biology" and noted that for this reason "the new ideas have. . .been assimilated and exploited faster in the geological sciences." But to any outside observer, aware of the radical change in our notions about the history of the earth, the magnitude of the shift in ideas would suggest a great revolution. Only because of the complete lack of an ideological component would it seem to be anything less.[11]

References for Chapter 10

Argand, Emile. 1977. *Tectonics of Asia*. Editor, Albert Carozzi. New York: Hafner Press.

Bullard, Edward C. 1965. "Historical introduction to terrestrial heat flow." Chapter 1 of *Terrestrial heat flow.* Washington, D.C.: American Geophysical Union [Geophysical Monograph no. 8].

____. 1975 "The effect of World War II on the development of knowledge in the physical sciences." *Royal Society of London Proceedings,* ser. A, 342: 519–536.

____. 1975a. "The emergence of plate tectonics: a personal view." *Annual Review of Earth and Planetary Sciences* 3: 1–30.

Bullen, K. E. 1976. "Wegener, Alfred." *Dictionary of Scientific Biography* 14: 214–217.

Cox, Allan. 1973. *Plate tectonics and geomagnetic reversals.* San Francisco: W. H. Freeman and Company.

Dietz, R. S. 1961. "Continent and ocean basin evolution." *Nature* 190: 854–857.

Du Toit, A. L. 1937. *Our wandering continents.* Edinburgh: Oliver and Boyd.

Frankel, Henry. 1984. "Biogeography before and after the rise of sea floor spreading." *Studies in History and Philosophy of Science* 15: 141–168.

Glen, William. 1982. *The road to Jaramillo: critical years of the revolution in earth science.* Stanford: Stanford University Press.

Greene, Mott T. 1982. *Geology in the nineteenth century: changing views of a changing world.* Ithaca: Cornell University Press.

Gutting, Gary, ed. 1980. *Paradigms and revolutions: applications and appraisals of Thomas Kuhn's philosophy of science.* Notre Dame, Ind.: University of Notre Dame Press.

Hallam, Anthony. 1973. *A revolution in the earth sciences: from continental drift to plate tectonics.* Oxford: at the Clarendon Press.

Harland, W. B. 1969. "Essay review: the origins of continents and oceans." *Geological Magazine* 106: 100–104.

Hess, H. H. 1962. "History of the ocean basins." Pp. 599–620 of A. E. J. Engel, H. L. James, and B. F. Leonards, eds., *Petrologic studies: a volume in honor of A. F. Buddington,* New York: Geological Society of America.

____. 1968. "Reply." *Journal of Geophysical Research* 73: 6569.

Holmes, Arthur. 1928. "Continental drift." *Nature* 122: 431–433.

Hurley, Patrick M. 1959. *How old is the earth?* Garden City, N.Y.: Doubleday.

Jeffreys, Harold. 1952. *The earth: its origin, history and physical constitution.* Cambridge: at the University Press.

Macrakis, Kristie. 1984. "Alfred Wegener: self-proclaimed scientific revolutionary." *Archives Internationales d'Histoire des Sciences* 112: 182–195.

Marvin, Ursula. 1973. *Continental drift: The evolution of a concept*. Washington: Smithsonian Institution Press.

McKenzie, D. P. 1977. "Plate tectonics and its relationship to the evolution of ideas in the geological sciences." *Daedalus* 106: 97–124.

Meyerhoff, A. A. 1968. "Arthur Holmes: originator of spreading ocean floor hypothesis." *Journal of Geophysical Research* 73: 6563–6565.

Opdyke, N. D. et al. 1966. "Paleomagnetic study of Antarctic deep-sea cores." *Science* 154: 349–357.

Simpson, George G. 1978. *Concession to the Improbable: an Unconventional Autobiography*. New Haven: Yale University Press.

Sullivan, Walter. 1974. *Continents in motion: the new earth debate*. New York: McGraw-Hill Book Company.

Takeuchi, H., S. Uyeda, and H. Kanamori. 1970. *Debate about the earth: approach in geophysics through the analysis of continental drift*. Rev. ed. San Francisco: Freeman, Cooper, and Co.

Totten, Stanley M. 1981. "Frank B. Taylor, plate tectonics, and continental drift." *Journal of Geological Education* 29: 212–220.

Uyeda, Seiya. 1978. *The new view of the earth: moving continents and moving oceans*. San Francisco: W. H. Freeman and Company.

Van der Gracht, Willem A. J. M. van Waterschoot, et al. 1928. *Theory of continental drift: a symposium on the origin and movement of land masses*. Tulsa: American Association of Petroleum Geologists.

Wegener, Alfred. 1905. *Die Alphonsinischen Tafeln für den Gebrauch eines modernen Rechners*. Inaug.-diss., Berlin.

____. 1915. *Die Entstehung der Kontinente und Ozeane*. Braunschweig: Friedrich Vieweg & Sohn.

____. 1924. *The origin of continents and oceans*. Trans. from 3rd German ed. by J. G. A. Skerl. London: Methuen.

____. 1966. *The origin of continents and oceans*. Trans. from 4th German ed. by John Biram. New York: Dover Publications.

Wegener, Else. 1960. *Alfred Wegener: Tagebücher, Briefe, Erinnerungen*. Wiesbaden: F. A. Brockhaus.

Wilson, J. Tuzo. 1963. "Continental drift." *Scientific American* 208: 86–100.

———. 1966. "Some rules for continental drift." Pp. 3–17 of G. D. Garland, ed., *Continental drift*, The Royal Society of Canada Special Publications, no. 9, Toronto: University of Toronto Press.

———. 1968. "Static or mobile earth: the current scientific revolution." *Proceedings of the American Philosophical Society* 112: 309–320.

———. 1968a. "Reply to V. V. Beloussov." *Geotimes* 13 (12): 20–22.

———. 1968b. "A revolution in earth science." *Geotimes* 13 (10): 10–17.

———, ed. 1973. *Continents adrift*. Readings from *Scientific American*. San Francisco: W. H. Freeman and Company. A second edition, 1976, entitled *Continents adrift and continents aground*.

Wilson, Leonard G. 1972. *Charles Lyell: the years to 1841: the revolution in geology*. New Haven, London: Yale University Press.

———. 1980. "Geology on the eve of Charles Lyell's first visit to America, 1841." *Proceedings of the American Philosophical Society* 124: 168–202.

Notes

Chapter 1

1. Letter from A. von Humboldt to Wilhelm Wegener, January 17, 1789.
2. Else Wegener provided me with a copy.
3. Abraham Gottlob Werner (1749–1815), famous German geologist and mineralogist: "father of geology" in Germany, taught after 1775 at the Mining Academy in Freiburg in Saxony. He believed granite and other igneous rocks to be deposits in the oceans ("neptunism").
4. *Die Jugendbriefe Alexander von Humboldts 1787–1799* (edited by Ilse Jahn and Fritz G. Lange), Akademie Verlag, Berlin, 1973. An older edition of W. G. Wegener's letters by Albert Leitzmann, 1896.
5. Georg Forster (1754–1794), German naturalist and writer, he accompanied James Cook on his second voyage; his account of the journey established him as one of the most advanced German thinkers of his time. In 1790 he travelled with A. von Humboldt along the Rhine River to Holland and England. He was an ardent supporter of the French Revolution; he died in poverty in Paris.
6. *Mineralogische Beobachtungen über einige Basalte am Rhein*, 1790.

Chapter 2

1. Translated from Latin into English by Greta Lindberg. Verse from "Gaudeamus igitur", a traditional student song.
2. A *Gymnasium* is a German secondary school whose curriculum emphasizes the classics, history, mathematics, and modern languages.
3. Kurt Wegener (born 1878, Berlin—died 1964, Munich), like his brother a meterologist; worked in Samoa, 1908; in Spitzbergen, 1912; at the German Marine Observatory in Hamburg, 1919–1923; took over his brother's Greenland expedition, 1931; professor in Graz, 1932–1941.
4. The Wegener family's burial site at the village church also has a stone for Alfred Wegener, although his grave is actually in Greenland.
5. Else Wegener, *Alfred Wegener*, 1960, p. 71.
6. Max Wolf (1863-1932), founded and directed the observatory on the Königstuhl near Heidelberg. See H.C. Freiesleben: *Max Wolf*, in the series *Große Naturforscher*, vol. 26.
7. Walter Lietzmann (1880–1939), *Aus meinen Lebenserinnerungen*, 1960. Also among Wegener's Berlin student friends was Walter Wundt (1883–1967), who later worked on Milankovitch's radiation curves and thus came into contact again with Wegener and Köppen.
8. Max Planck (1858–1947), Nobel Laureate in Physics, 1918; Emil Fischer (1852–1919), Nobel Laureate in Chemistry, 1902.
9. Wilhelm von Bezold (1837–1907), after 1885, Professor in Berlin and Director of the Prussian Meteorological Institute. Essentially, however, Wegener was self-taught in meteorology, as August Schmauss points out in a memorial essay (*Annalen der Meterologie* 4, 1–13, 1951).
10. J. Bauschinger (1860–1934); Professor of Astronomy in Berlin, 1896–1909; then in Leipzig.
11. Wilhelm Foerster (Wegener writes "Förster"), 1832–1921, wrote an autobiography, *Lebenserinnerungen und Lebenshoffnungen*

(1911; Wegener is not mentioned in it). Foerster founded the "Urania" Society and its observatory, which were intended to educate the public. Today the observatory on the "Insulaner" hill in Berlin bears his name.

12. Alfonso X of Castille, "the Wise" (1221–1284).
13. Quoted in F. Neumann, *Nicolaus Copernicus* (Münster, 1973).
14. August Schmauss, *Annalen der Meteorologie* 4, 1–13, 1951.
15. The Royal Prussian Aeronautic Observatory was moved from Berlin to Lindenberg in 1904; directed by Richard Assmann (1899–1914). Assmann developed an aspiration psychrometer to measure humidity; he was one of the discoverers of the stratosphere (1902).
16. Franz Richarz (1860–1920), after 1901, professor of physics in Marburg.
17. Hans Cloos (1885–1951), professor of geology in Breslau and Bonn. Quote from his book *Gespräch mit der Erde*, 1947.
18. Emanuel Kayser (1845–1927), at the University of Marburg from 1885–1917.
19. Johannes Georgi (1888–1972), meteorologist; studied in Marburg under Wegener; after 1919, at the German Marine Observatory in Hamburg; with Wegener in Greenland 1929 and 1930–1931, wintered at "Mid-Ice" 1930–1931. Memorial essay by F. Loewe in *Polarforschung*, 1972, pp. 155–158.
20. Vilhelm Bjerknes (1862–1951) taught at the University of Leipzig, 1912–1917.
21. Johannes Georgi, "Memories of Alfred Wegener," in S. K. Runcorn (editor), *Continental Drift*, Academic Press, 1962, pp. 309–324. Included in this book as Chapter 9.
22. The German Marine Observatory in Hamburg (founded 1875) grew out of the North German Marine Observatory. Its first, energetic director was Georg von Neumayer. In 1945 the work of the German Marine Observatory was taken over by the German Hydrographic Institute and the Marine Weather Office of the German Weather Service.

23. Franz Grillparzer, March 25, 1819, an Austrian poet (1791–1872).
24. Heinrich von Ficker (1881–1957), director of the Prussian Meteorological Institute, 1923–1934; after 1937, director of the Central Institute for Meteorology in Vienna.
25. Victor Franz Hess (1883–1964), professor in Graz, 1920–1931 and 1937; after 1938, in the United States; he received the Nobel Prize for Physics in 1936 for studies of cosmic radiation.
26. 1931, *Gerlands Beiträge zur Geophysik* 31, 337–377.

Chapter 3

1. Alexander Pushkin (1799–1837), the famous Russian poet, quoted in P. Gordijenko, *Die Polarforschung in der Sowjetunion*, 1967, p. 67.
2. Hans Benndorf, 1931, *Gerlands Beiträge zur Geophysik* 31, 337–377.
3. Achton Friis, *Mit Mylius-Erichsen im Grönlandeis*. German edition, Leipzig, 1910. See also A. Wegener, 1911, 1912, and J. Georgi in *Continental Drift*, edited by S. K. Runcorn, 1962.
4. J.P. Koch, *Durch die weiße Wüste* (translated into German by A. Wegener, 1919). —Also, A. Wegener, *Tagebuch eines Abenteuers (Diary of an Adventure)* (editor E. Wegener), 1961. Georgi objected to E. Wegener's choice of the word "Abenteuer" (adventure) for the title of the posthumously published book (*Polarforschung*, V, 1961 (1963), p. 123). But the title does not make Wegener himself an adventurer, and there was certainly adventure enough on the Greenland expeditions; one need only reread the first sentence of the letter Georgi himself wrote from "Mid-Ice" (see Chapter 3)!
5. A. Wegener, *Mit Motorboot und Schlitten in Grönland*, 1930, and "Deutsche Inlandeis-Expedition nach Grönland Sommer 1929." *Zeitschrift der Gesellschaft für Erdkunde Berlin* 1930, 81–124.

6. Fritz Loewe (1895–1974), meteorologist. Memorial essay by Karl Weiken, *Polarforschung* 44, no. 1, 1974, pp. 92–95.
7. Ernst Sorge (1899–1946), also went on 1930–1931 expedition.
8. Else Wegener, editor, *Alfred Wegeners letzte Grönlandfahrt*, 1932. See also Wegener's bibliography.
9. *Polarforschung*, 40, 1970, p. 2–3 (original text in French). Quote from the French explorer Paul-Emile Victor, who reached the site of "Mid-Ice" in 1949.
10. Kurt Wegener, editor, *Wissenschaftliche Ergebnisse der Deutschen Grönland—Expedition* I, 1933, p. 35.
11. Journal entries taken from Else Wegener, *Alfred Wegener*, 1960, and Else Wegener, editor, *Alfred Wegeners letzte Grönlandfahrt*, 1932, and in part from the originals in the Deutsches Museum, Munich. Else Wegener occasionally made slight changes in the journal entries. For example, in the entry shown in the figure she inserted the phrase "fully loaded" after "motorized sleds" (Else Wegener, *Alfred Wegener*, 1960, p. 244). While the emendations sometimes aid understanding, they are of course made at the expense of authenticity. See also Georgi, 1960.
12. Wegener is referring to a motor boat used by the expedition, which had been brought to Greenland in 1929.
13. Else Wegener, *Alfred Wegener*, 1960, p. 238.
14. This letter, written in pencil, was acquired in 1978 from the autograph dealer Stargardt-Marburg. I am grateful to Dr. M. Rodewald and Frau Dr. B. Seiler, Hamburg, for identifying the recipient.
15. Else Wegener, editor, *Alfred Wegeners letzte Grönlandfahrt*, 1932 pp. 185–186.
16. F. Schmidt-Ott describes the reception of the expedition in Copenhagen on November 13, 1931, in his memoirs (*Erlebtes und Erstrebtes*, 1952): "On the previous evening I had already had a two-and-a-half hour conversation with Dr. Georgi und Dr. Sorge, who shook hands with me on a peaceable solution. Following a reception and dinner in the Hotel d'Angleterre, to which I had invited twenty-two guests, there was a meeting of the expedition

members, at which Else Wegener was also present. Calling on the memory of Alfred Wegener, I succeeded in concluding peace among them."

In his later official final report, published by the *Notgemeinschaft* in 1933, Kurt Wegener found no cause to criticize Georgi's conduct in any way. Rather, he reported objectively on Georgi's scientific contributions. If, nevertheless, Else Wegener mentions Georgi again, years later in her biography of her husband, this can hardly be seen as a condemnation of Georgi's conduct. Hers is not the criticism of an expert member of the expedition or the *Notgemeinschaft*—its judgment had gone in Georgi's favor long ago—but rather it is a woman's anxious question, directed less at the head of the "Mid-Ice" station than at inexorable fate. Perhaps one can understand why Georgi, in spite of everything, reopened the unfortunate "question of guilt" again in 1960. Less easy to understand is his stinging and often petty criticism of Else Wegener's biography. That earned him no good will, and was altogether unfortunate, given his undeniably great contribution to the expedition. —See also H. Beck in his short Wegener biography of 1971 (*Grosse Reisende*, Munich, pp. 314–330). But even as recently as 1978, in Büdel's biography (*Die Grossen der Weltgeschichte*, Kindler-Zurich, 11, pp. 460–467), one finds yet again the (unfounded) claim that "others" contributed to Wegener's death.

17. *Petermanns Geographische Mitteilungen* 77, pp. 169–171, 1931.

18. In some respects, one could compare Wegener's achievement, with the dog-sled journey in 1978 of a Japanese who covered the 800 kilometers between Ellesmere Land and the North Pole in eight weeks. Through radio contact with a supply plane, provisions were replenished and the sled dogs changed as necessary. The nonetheless arduous undertaking had no scientific objectives whatsoever. (*National Geographic* 154, Sept. 3, 1978).

19. The remains of the expedition's two motorized sleds were taken to the Danish Technical Museum in Helsingör in 1973 (A. Weidick, *Polarforschung* 44, 1, 1974). On August 29, 1930, they had been christened "Snowbird" and "Polar Bear". "Weiken per-

formed the baptism. Snowballs took the place of champagne bottles" (Journal, August 29).

20. A. Staxrud and K. Wegener. *Die Expedition zur Rettung von Schröder-Stranz und seinen Begleitern*. Published by A. Miethe, Berlin, 1914. Of particular note is Kurt Wegener's critical remark (p. 101) on the tendency of those who remain at home to form "hasty, superficial, and ill-considered judgments."
21. *Meteorologische Zeitschrift* 48, 1931.
22. *Annalen der Meteorologie* 4, 1–13, 1951.
23. It has sometimes been assumed that Wegener's objective in the 1930 expedition was to verify directly through measurements that Greenland is drifting. If so, his death would have been a tragic consequence of his hypothesis (see, for example, P. Jordan, *Expansion der Erde*, 1966, p. 52; P. Schmidt-Thomé in *Lehrbuch der Allgemeinen Geologie*, II, 1972, p. 514; H. G. Wunderlich, in *Das neue Bild der Erde*, 1975, p. 305). Actually, Wegener's major goals for this expedition were climatological, as is evident from the expedition's plan (Else Wegener, *Alfred Wegener*, 1960, and *Polarforschung*, 1960, Supplement 2, p. 45ff.); see also J. Georgi in *Polarforschung* 1960, Supplement 2, p. 13.

Chapter 4

1. Translated from Icelandic into English by Professor Richard Ringler. From a poem by the Icelandic poet Grimur Thomsen (1820–1896) called "In the Sprengisandur Desert." The poem captures the anxiety of a horseman crossing the desolate central highland of Iceland, which was traditionally believed to inhabited by trolls and elves.
2. J.P. Koch, "Die Reise durch Island 1912." *Petermanns Geographische Mitteilungen*, October 1912.
3. I am grateful to Ulfur Fridriksson of Reykjavik for valuable information about the Icelandic "travelers to Greenland."

4. Niels Nielsen, born 1893, was professor of geography in Copenhagen from 1939 to 1964.
5. Volcanoes sometimes form over long fissures which have served as conduits for lava. Among such "fissure eruptions" in Iceland were, for example, the "Laki" eruptions of 1783 and the first stages of the eruption on Heimaey in 1973.
6. Otto Niemczyk (1886–1961), born in Upper Silesia; worked as a mining surveyor; after 1931, professor at the Technische Hochschule in Berlin.
7. Ferdinand Bernauer (1892–1945), petrographer and volcanologist at the Technische Hochschule in Berlin.
8. Personal communication from Professor K. Gerke, Dec. 1978.
9. K. Gerke et al., "Geodätische Lagemessungen zur Bestimmung horizontaler Krustenbewegungen in Nordost-Island." *Festschrift Walter Höpcke*, Hannover, 1978.
10. K. Saemundsson, "Fissure swarms and central volcanoes of the neovolcanic zones of Iceland." In: D. R. Bowes and B. E. Leake, "Crustal evolution," *Geologic Journal*, Special Issue, 10, 1978. —For the opposing view see, for example, V. V. Beloussov and E. E. Milanovsky, "On tectonics and tectonic position of Iceland," *Greinar V. Visindafæelag, Island,* Reykjavik, 96–118, 1976.

Chapter 5

1. Georg Christoph Lichtenberg (1742–1799). German physicist and writer.
2. Journal of the "Danmark" expedition, February 1, 1907.
3. *Annalen der Meteorologie* 4, 1951, 1–13.
4. Alexander I. Woeikoff (or Voeikov) (1842–1916), prominent Russian climatologist. Else Wegener remembered him as a highly eccentric guest in her family's home: "In spite of his wealth, he neglected his personal appearance terribly. No wonder no hotel

in Hamburg wanted to take him in. He lived with us, and my mother would surreptitiously mend his clothes." (Biography of W. Köppen, 1955, p. 102.)

5. Halo effects, rings, and other similar optical effects around the sun and moon are created by refraction through ice crystals.
6. Professor R. Huckriede of Marburg kindly provided me with the selection from E. Kayser's memoirs (written in 1927) on the meteorite of Treysa (by permission of Luise Kayser).
7. In the book by N.M. Short, *Planetary Geology* (Prentice Hall Inc.), 1975, which contains a section devoted to "experimental cratering", they are not mentioned, nor are they in other contemporary books on the moon. In his many articles on "selenology", K. von Bülow very seldom cites Wegener.
8. Ellen Drake, "Alfred Wegener's reconstruction of Pangaea," *Geology* 4, 4, 1976, pp. 41–48. —An often quoted computer study on the best possible fit of the continents on the two sides of the Atlantic was done by E. Bullard, J. E. Everett, and A. G. Smith, *Philosophical Transactions Royal Society London* A, 258, 1965; a similar reconstruction was made by S. W. Carey, *Continental Drift-Symposium*, Hobart (Tasmania) 1958. p. 218ff —For the history of the drift hypothesis see also the compilation *Geologie und Paläontologie in Texten und ihrer Geschichte* by H. Hölder, Freiburg and Munich; among English-language studies, see A. Hallam, *A Revolution in the Earth Sciences. From Continental Drift to Plate Tectonics* (Oxford, 1973). Also see A. V. Carozzi, "A propos de l'origine des dêrives continentales." *Société de Physique et d'Histoire Naturelle, Compte Rendu des Séances*, Genêve, N.S. 4, 171–179, 1969; W.B. Harland, "The origin of continents and oceans: an essay review." *Geological Magazine* 106, 100–104, 1969; N. A. Rupke, "Continental drift before 1900." *Nature* 227, 349–350.
9. *Nature* 126, 1930, p. 841.
10. W. H. Pickering: *Journal of Geology* 15, 23–38, 1907; H. B. Baker: *Michigan Academy of Sciences, Annual Report*, 1912, 1913; F. B. Taylor: *Bulletin of the Geological Society of America* 21,2, 1910.

11. *A Revolution in the Earth Sciences: From Continental Drift to Plate Tectonics*. Oxford University Press.
12. Since the end of the nineteenth century, the view has prevailed that in the Alps entire blocks of mountain range were overthrust over other strata from the south.
13. Eduard Suess (1831–1914); after 1857, professor of geology in Vienna. His major work was *Das Antlitz der Erde*, Vol. I–III, (1885–1909). Influential advocate of the "contraction theory"—creation of ranges of folded mountains through contraction of the earth as it cooled. "It is the collapse of the globe that we are experiencing."
14. Robert Schwinner (1878–1953), professor of geology in Graz.—Biography by H. Flügel, *Publikationen aus dem Archiv der Universität Graz* 7, 1977.
15. Quoted in E. Wegener, *W. Köppen*, 1955, p. 102–103.
16. *Meteorologische Zeitschrift* 38 (1921), 4, 97–101. However, the term "paleoclimatology" is not completely new; there are isolated instances of its use in the nineteenth century.
17. Henry Potonié (1857–1913), geologist at the Prussian Geological Institute in Berlin; paleobotanist.
18. *Durch ferne Welten and Zeiten*, 1936.
19. At first, "ice age" climatic conditions were known only for the epoch which immediately preceded our own—the Pleistocene. The Pleistocene and the chronologically insignificant present epoch make up the Quaternary Period, which began about three million years ago.
20. Albrecht Penck (1858–1945), did important research on the Quaternary Ice Age; professor in Vienna and, after 1906, Berlin.
21. Eduard Brückner (1862–1927), first taught physical geography at the German Marine Observatory in Hamburg (he was Else Wegener's godfather); after 1906, professor in Vienna (Penck's successor).
22. The cause of the variations in solar radiation to the earth lies in the periodic variation of elements of the earth's orbit: the angle of the earth's axis (between 24° 36′ and 21° 58′) in 20,000

years; the eccentricity of the orbit in 95,000 years; and the complete cycle of the vernal equinox in 21,000 years. The variations in radiation can be calculated precisely for any latitude or place on the earth's surface and can be plotted as "radiation curves."

23. J. Croll, *Climate and Time in Their Geological Relations*, London, 1875 (and later editions).
24. Milutin Milankovitch (1879–1958); after 1909, professor of astronomy in Belgrade, Yugoslavia. He published the first detailed account of radiation curves in Paris in 1920; a later account appeared in the *Handbuch der Klimatologie I, A*, 1930 (*Mathematische Klimalehre und Astronomische Theorie der Klimaschwankungen*). More recent calculations—differing from Milankovitch's results only in some small points—were done by A.J.J. van Woerkom (1953) and A.D. Vernekar (1972) in the United States.
25. *Handbuch der Geophysik* IX, 3, 1938, p. 690. —I. Schaefer (1953); *Geologica Bavarica* 19, 7–12, is in error when he gives priority for the geological application of the radiation curves to the Bavarian geologist B. Eberl. That is clear from Milankovitch's statements (see also note 18) and from Eberl's own statements *Zeitschrift der Deutschen Geologischen Gesellschaft* 80, 1928, Mber., p. 115; also *Die Eiszeitenfolge im Nördlichen Alpenvorderland*, Augsburg, 1930, p. 379–380). Soergel, too, came *after* Köppen and Wegener.
26. Wolfgang Soergel (1887–1946), geologist, professor in Breslau and Freiburg im Breisgau. —Friedrich Zeuner (1905–1963), in England after 1934.
27. Richard Foster Flint (1902–1976), professor at Yale University, New Haven; in both senses of the word "one of the giants of geology" (R. W. Fairbridge in an obituary in *Eiszeitalter und Gegenwart* 27, 1976).
28. Parkin, *Nature* 260, 1976. The study by Hays et al. appeared in *Science* 194, Dec. 1976, 1121–1132.
29. The Köppen quote is from *Grundriß der Klimakunde* (1931, p. 43). The meteorologist H. Flohn, among others, has often expressed his opinion on future climatic developments; see also

K. Duphorn in "Kommt eine neue Eiszeit?" *Geologische Rundschau* 65, 845–864, 1976).

30. Rodolphe Toepffer, Swiss writer, *Nouvelles Gênevoises*, 1841.
31. "Medium" duration, several tens of thousands to 100,000 years; "long" duration, several tens of millions to 100 million years. See also Schwarzbach, *Klima der Vorzeit*, 3rd ed., 1974, p. 266.
32. *Annals of the New York Academy of Sciences* 95, 1961, p. 569.
33. *Autobiography*, 1876. Charles Darwin made this remark, which was not meant seriously, to his friend Charles Lyell (1797–1875) many years previously (before 1876). Darwin specifically points out that Lyell, who was for many years a committed opponent of the theory of natural selection, "became a convert. . .when he had grown old."

Chapter 6

1. *Scientific Autobiography*, 1945. This quote from Max Planck on "a new scientific truth" is relevant to the hypothesis of continental drift. It is true that Wegener's explanation of the transport mechanism was incorrect, but his hypothesis of large-scale displacements seems to be "a scientific truth." (See also the final section of Chapter 8). In a stimulating discussion in 1977 of the evolution of theories in the modern earth sciences, W. von Engelhardt cites continental drift as an example.
2. Geophysics—"in the narrower sense, the physics of the solid earth. Geophysics is concerned with gravity, the seismic, thermal, magnetic, and electric phenomena of the earth, and the physical structure of its interior." *Geologisches Wörterbuch* by H. Murawski, 7th ed., 1977.
3. Gustav Steinmann (1856–1929), professor of geology in Freiburg im Breisgau and Bonn.
4. In *Gespräch mit der Erde*, p. 364. See also Chapter 2, Marburg.

5. Hermann von Ihering (1850–1930), zoologist and geologist; in South America from 1880 to 1920; after 1926, honorary professor of zoology and paleontology in Gießen.
6. Fritz Kerner-Marilaun (1866–1944), son of the Alpine botanist Anton Kerner-Marilaun; geologist at the Imperial Geological Institute in Vienna; author of *Paleoclimatology* (1930).
7. Max Semper (1870–1952), grandnephew of the architect Gottfried Semper; professor of geology in Aachen from 1924 to 1935.
8. Wilhelm Salomon-Calvi (1868–1941), professor of geology in Heidelberg; emigrated to Turkey in 1934.
9. The word "epeirophoresis" is derived from the Greek word *pherestai*, "being driven" (Salomon-Calvi, *Sitzungsberichte der Heidelberger Akademie der Wissenschaften*, 1930).
10. Hans Stille (1876–1966), professor of geology in Göttingen and Berlin.
11. *Zeitschrift der Deutschen Geologischen Gesellschaft* 128, 1977.
12. *Geology* 6, Oct. 1978, p. 586.
13. Hans Georg Wunderlich (1928–1974), *Das neue Bild der Erde*, Hamburg, 1975.
14. Robert Scholten, *Geology* 6, Oct. 1978, p. 586.
15. Ellen Drake (1976, p. 41), Frederic Vine (1977, p. 20), and Robert Scholten (1978) erroneously assumed that Wegener was present—perhaps because Wegener did submit a paper which appeared in the proceedings of the conference, which were published in Tulsa, Oklahoma, in 1928. The term "Tulsa Symposium", which crops up occasionally in the literature, is also erroneous. The meeting was held in New York: the proceedings were published in Tulsa.
16. *Nature* 266, 1977, 19–22.
17. *American Journal of Science* 242, 1944, 510–513.
18. P. Jordan, *Die Expansion der Erde: Folgerungen aus der Diracschen Gravitationshypothese*. Braunschweig, 1966, p. 66.
19. Arthur Holmes (1890–1965), ended his career as professor of geology in Edinburgh; author of an excellent textbook called

Principles of Geology. He wrote about the New York Symposium in *Nature*, 122, 1928, Sept. 22.

20. *American Journal of Science* 242, 1944, p. 514–515.
21. *Nature* 171, 669–671, 1953. —The remark on "sterility" appears in *Principles of Physical Geology*, 1965, p. 1203.
22. Otto Ampferer (1875–1947), ended his career as the director of the Federal Geological Institute in Vienna. He was a "leading Alpine geologist of unique character" (W. Heißel). In his study *Über das Bewegungsbild von Faltengebirgen* (1906) he described his "theory of undercurrents." The most recent appreciation of Ampferer's work appeared in *Beiträge zur Technikgeschichts Tirols* 7, Innsbruck, 1977 (by W. Heißel).
23. (1878–1953), professor of geology in Graz.
24. Original in the Deutsches Museum in Munich.
25. From *The Wandering Continents* by Alexander Du Toit, 1937.
26. A. L. Du Toit (1878–1948). Obituary note by S. H. Haughton *Proceedings of the Geological Society of America for 1949*, 141–149).
27. Sir Edward Sabine (1788–1883), astronomer; on his scientific expeditions he established meteorologic-magnetic observatories and used pendulums to investigate the shape of the earth; he also recognized the connection between sun spots and magnetic disturbances.

Chapter 7

1. "Eppur si muove." Galileo was referring to his belief that the earth revolves around the sun.
2. Emile Argand (1879–1940), professor of geology in Neuchatel. *La tectonique de l'Asie* appeared in the *Transactions of the 13th International Geology Congress* (Brussels), I, 5, 171–372, 1922 (English translation by A.V. Carozzi), 1977; reviewed by J.F.

Dewey in: *Geology* 6, March 1978, and by R. Scholten in: *Geology* 6, Oct. 1978.

3. *Geology* 6, Oct. 1978, p. 586.
4. *Continental Drift*, ed. S.W. Carey, 1958, p. 10.
5. Asthenosphere—formed from the Greek words "asthenes" (weak) and "sphaira" (sphere). The asthenosphere probably lies between 100 and 300 kilometers below the surface.
6. Wegener believed that the "Atlantic fissure" did not begin to open up significantly until the Quaternary period.
7. "Der Begriff der Konvektionsströmung in der Mechanik der Erde", *Gerlands Beiträge zur Geophysik* 58, 1941, 119–158, p. 153. Schwinner was able to base his theory on the detailed catalogs of deep focus earthquakes that B. Gutenberg and C.F. Richter had assembled (*Bulletin of the Geological Society of America* 49 and 50, 1938 and 1939). However, the two California seismologists had drawn no conclusions about the sloping zones of movement.
8. It was not only the Pangaea of the late Paleozoic that split apart; the modern continents also show signs of splitting—such as large graben zones (the "Rhine graben" between the Black Forest and the Vosges; the African lakes, etc.). Thus the formation of intra- or microplates (although they are, of course, still quite large) is possible. See for example, H. Illies, *Geologische Rundschau* 64, 1975, 677–699.
9. Forty-five papers, most of which deal with this question, are collected in the volume *Mineral Deposits, Continental Drift and Plate Tectonics* (editor, J. B. Wright, Benchmark Papers, 44, 1977).
10. *Erzmetall* 24, 6, 257–306, 1972.
11. Pierre Termier (1859–1930), important French geologist. "The theory of Wegener is to me a beautiful dream, the dream of a great poet. One tries to embrace it and finds that he has in his arms but a little vapor or smoke; it is at the same time both alluring and intangible." *Annual Report of the Smithsonian Institution for 1924*, Washington, 1925, p. 236.

12. M.H. Nitecki et al., Acceptance of plate tectonic theory by geologists. *Geology* 6, 11, 1978.
13. "Continental drift. New orthodoxy or persuasive joker?" In: D. H. Tarling and S. K. Runcorn, *Implications of Continental Drift* II, 1973.
14. For example, W. Krebs in 1973 (*Geologic Society of America Bulletin* 83, 8, 2611-2630, with H. Wachendorf) and in 1975 (*American Association of Petroleum Geologists* 59, 9, 1637–1666); H. G. Wunderlich in 1975; Zeil in 1975; with regard to paleoclimatology see for example H. Kozur *(Nova Acta Leopoldina N.F.* 224, 45, 413–472, 1976). Meyerhoff and Beloussov were already mentioned in Chapter 4 (Iceland).
15. A constant expansion of the earth would explain why the continents have moved apart over the course of the earth's history, without need for the drift hypothesis. But it is very difficult to find any sort of reason for such an expansion, although physicists have discussed it as a theoretical possibility (Pascual Jordan, *Die Expansion der Erde*, 1966). The geologist S.W. Carey (University of Tasmania, Hobart), today the main advocate of an "expanding earth", answered his own question about the reason for expansion: "My first answer is I do not know. Empirically I am satisfied that the earth is expanding." (*The Expanding Earth*, Amsterdam, 1976, p. 446). As early as 1956 Carey had organized a symposium on continental drift in Hobart. One of the first to attribute the cause of continental drift to expansion was Otto Hilgenberg in his detailed presentation of the idea: *Vom wachsenden Erdball* (Berlin, 1933). See also his study "Paläopollagen der Erde" (*Neues Jahrbuch für Geologie und Paläontologie*, Abhandlung 116, 1–56, 1962). Among geophysicists, the Hungarian Laszlo Egyed is an advocate of the theory.
16. Werner Zeil, "Allgemeine Geologie". *Brinkmanns Abriß der Geologie* I, 11th ed., Stuttgart (Enke), 1975.
17. *Nature* 266, March 3, 1977, p. 22.
18. Sir Arthur Eddington (1882–1944), English astronomer and philosopher; after 1913, professor in Cambridge. —*Nature* 111, 1923, 18–21.

19. *Proceedings of the Geological Society of London*, 1957, p. 79.
20. Dialogue on the two major views of the solar system. Leipzig, 1891, p. 59. Quoted in J. Wickert, *Einstein*, RoRoRo Bildmonographie, 1972, p. 34.

Chapter 8

1. *Scientific American* 232, no. 2, 1975, p. 95.
2. *Nature* 266, March 3, 1977, p. 22
3. *A Revolution in the Earth Sciences: from Continental Drift to Plate Tectonics*, Oxford University Press, 1977, p. 105.
4. The importance of the new ideas in America, can be ascertained, for example, from a glance into a modern American textbook of physical geology. (F. S. Sawkins et al., 2nd ed., 1979), in which almost one-third of the book is devoted to plate tectonics and associated processes. That is in contrast to the slightly older German textbook *Tektonik* by P. Schmidt-Thomé (Stuttgart, 1972), where only 15 out of 579 pages deal with plate tectonics. In large, multi-volume German dictionaries, plate tectonics began to appear only in 1974.
5. *Nature*, February 16, 1922.
6. *Proceedings of the American Philosophical Society* 112, 5, Oct. 1968.
7. "Adalbert von Chamisso as a scientist." Memorial address, Prussian Academy of Sciences, Berlin, 1888. In the major biography of Humboldt by Hanno Beck (1961; with one chapter entitled "Profile of a Genius") this interesting quotation is not cited.

Chapter 9

1. H. Benndorf, *Gerlands Beiträge zur Geophysik* 31, 337, 1931.
2. S. K. Runcorn, *Science* 129, 1002, 1959.
3. F. Rossmann, *Zeitschrift für angewandte Meteorologie* 48, 257, 1931.
4. A. Wegener, *Petermanns Geographische Mitteilungen* 58, pp. 185, 253, 305, 1912.
5. S. Saxow, *Medd. dansk geol. Foren.* 13, 522, 1958.
6. W. Köppen, *Petermanns Geographische Mitteilungen* 77, 169, 1931.
7. Else Wegener, *Alfred Wegener*, Wiesbaden, 1960.
8. Various authors. In *Meddelelser om Grønland* 42 and 46, 1909 and 1911.
9. Achton Friis, *Im Grönlandeis mit Mylius Erichsen; die Danmark-Expedition 1906–08*, Leipzig, 1910.
10. A. Wegener, *Thermodynamik der Atmosphäre*, Leipzig, 1911.
11. G. H. Darwin, *Tides and Kindred Phenomena in the Solar System*, London and Boston, 1898; and German edition: *Ebbe und Flut usw.*, Leipzig, 1902.
12. J. P. Koch and A. Wegener, *Durch die weisse Wüste, die dänische Forschungsreise quer durch Nordgrönland 1912–13*, Berlin, 1919. For the scientific report see note 15.
13. W. Köppen and A. Wegener, *Die Klimate der geologischen Vorzeit*, Berlin, 1924.
14. E. Wegener and E. Kuhlbrodt, *Wladimir Köppen, ein Gelehrtenleben*, Stuttgart 1955. Includes a list of Köppen's publications.
15. J. P. Koch and A. Wegener, *Meddelelser om Grønland*, 75, Copenhagen, 1928.
16. J. Georgi, *Im Eis vergraben. Erlebnisse auf Station Eismitte der letzten Grönland-Expedition Prof. A. Wegener's 1930–31*, p. 303, Leipzig, 1955. Material on the early history of the Mid-Ice Station.
17. A. Wegener, "Denkschrift über Inlandeis-Expedition nach Grönland" *Deutsche Forschung (Aus der Arbeit der Notg.d.D.Wiss.)* H.21, 181. Berlin, 1928; reprinted in note 19. This plan was to

form the basis of Wegener's preparations for and execution of the 1929 and 1930–31 expeditions. The plan as published in note 18 was not authorized.

18. Wissenschaftliche Ergebnisse der Deutschen Grönland-Expedition, Vol. 1, 3, Leipzig, 1933.
19. J. Georgi, *A Wegener zum 80. Geburtstag (1.11.60). Polarforschung* 2, Supplement 1960, 45.
20. R. A. Hamilton, *Polarforschung* 28, 103, 1958.
21. A. Wegener, *Mit Motorboot und Schlitten in Grönland*, Leipzig, 1930.

Chapter 10

Reprinted by permission of the publishers from *Revolution in Science* by I. Bernard Cohen, Cambridge, Mass.: The Belknap Press of Harvard University Press, Copyright © 1985 by the President and Fellows of Harvard College.

1. Wegener did admit that in some aspects Taylor's "viewpoint. . .differs only quantitatively from my own, but not in crucial or novel ways." He observed that "Americans have called the drift theory the Taylor-Wegener theory," but he insisted that "in Taylor's train of thought continental drift in our sense played only a subsidiary role and was given only a very cursory explanation" (1966, 4).

   Wegener's earliest paper on continental drift was published in 1912. In it he referred (pp. 185, 194–195) to Taylor's earlier publication of 1910. Years later, in the fourth edition of his book (1929), Wegener wrote up a history of the development of his ideas, claiming that he had never heard of any other work on this subject until after he had himself conceived of the main ideas of the hypothesis of continental drift. In particular, he alleged (p. 3) not to have heard of Taylor's ideas. Wegener said that he formulated his ideas during the months from autumn 1911 to early January 1912. Taylor's paper of 1910 was published in June, but had been read at a meeting of the Geological Society of America in December 1908. A review of Taylor's published paper, by Jesse Hyde, appeared in the *Geologische Zentralblatt* for 15 April 1911, which should have attracted Wegener's attention.

In a recently discovered letter by Taylor to the editor of *Popular Science Monthly*, written on 4 December 1931, it is claimed (see Totten 1981, 214) that Wegener published "a very brief note" in "the spring of 1911," reviewing Taylor's paper, "partly in terms of approval, but putting forth some suggestions of his own." Taylor had either "mislaid or lost" his copy of the note, which he had "not been able to find. . .for three or four years." He had tried unsuccessfully, he wrote, to find this article in "the German scientific journals in the library of the University of Michigan at Ann Arbor." Harold Totten, who edited and published Taylor's letter, reported that "attempts by me and others to uncover the note have been unsuccessful."

Taylor described this note, which he says Wegener published in 1911, as "about twenty or twenty-five lines long," in "fine type." It is possible that Taylor's memory was at fault, and that the published report he had in mind was not by Wegener at all, but was the review by Jesse Hyde in that same year 1911, and which Taylor did not cite in his letter.

In an analysis of the texts of Taylor and Wegener, Totten (1981) finds strong evidence of Taylor's likely influence on Wegener. Not only are there "many similarities," but "in the case of Greenland, the reconstructions are nearly identical." Totten finds Taylor "justified in his claim" that "Wegener got at least some of his ideas" from him. What is perhaps of equal interest is the fact that Taylor had an early concept of plate tectonics, moving crustal sheets which, in collision, formed Tertiary mountain belts.

2. Wegener's thesis (1905) consisted of a modernization of the "Alfonsine Tables" of planetary motion, which were named after King Alfonso el Sabio (1221–1284) of Castille, who sponsored a redaction and new edition in Spanish of the "Toledan Tables" of the Cordoban astronomer al-Zarqālī or Arzachel. Wegener con-

verted the old sexagesimal numbers in the original tables into current decimal numbers, so that today's astronomers and chronologists could have more ready access to them in making their computations. The thesis was entitled *The Alfonsine Tables for the Use of Modern Computers*, which in 1905 meant 'for the use of men and women doing computations'. Wegener was deeply interested in the history of astronomy, but at that time no doctorates were being awarded in the history of science and there were no academic posts for specialists in the history of astronomy (Marvin 1973, 66). Wegener's edition of the Alfonsine Tables for modern users has proved to be a valuable tool for twentieth-century astronomers needing numerical planetary and lunar data of an earlier period, and there has been a recent reprint. But today's historians of astronomy prefer to use the originals, in their sexagesimal notation, and consider Wegener's efforts to have been misguided.

3. Hallam (1973, 9) has observed that "the subsequently coined term *continental drift* caught on universally in the English-speaking world presumably because people would rather not utter seven syllables when five will suffice."
4. Thus, R. T. Chamberlin of the University of Chicago: can geology be considered a science if it is "possible for such a theory as this to run wild?" (p. 83). In a review of the published proceedings in *Geological Magazine* for September 1928, "P. L." (presumably Philip Lake) stressed the unfavorable comments, quoting with approval R. T. Chamberlin's remarks that the "appeal" of the Wegener hypothesis "seems to lie in the fact that it plays a game in which there are few restrictive rules, and no sharply drawn code of conduct. So a lot of things go very easily." Bailey Willis of Stanford University said simply that Wegener's book was "written by an advocate rather than by an impartial investigator." According to Edward W. Berry, Wegener's method "is not scientific"; Wegener ends up "in a state of auto-intoxication in which the subjective idea comes to be considered as an objective fact."
5. The analogy between Wegener's theory and the Copernican theory (and Galileo's condemnation for advocating it) was a familiar

theme of the 1920s (used by Daly [1926] and Chamberlin [1928] and revived by many writers in the 1960s and 1970s.

6. In the 1928 conference, Schuchert—after taking 41 pages to show that there are serious misfits in putting together the continental "jig saw puzzle" and to question the paleontological evidence used by Wegener, challenging him on methodological grounds—nevertheless concluded fairly by contrasting what he called his own "iconoclastic" position toward "the Wegener hypothesis" with his "open-minded" stance on "the idea that the continents may have moved slowly, very slowly indeed, laterally, and differently at different times" (p. 141). He thus was willing to go along with Daly and Emile Argand in accepting the general idea of mobility and, after reminding his readers that land masses must indeed have moved on the earth, he said that one "begins to remember the statement of Galileo in regard to the earth: 'And yet it does move.' " In thus going along with the general idea of mobilism, but nevertheless rejecting the specific form of mobilism proposed by Wegener, Schuchert admitted that he had been influenced by Daly, whose just published book "sounds the keynote for the attempt to save the germ of truth in the displacement theory and reconcile it with the facts that geology already has at hand" (pp. 142, 144).
7. This letter was found by Kristie Macrakis among Wegener's Nachlass, in the Library of the Deutsches Museum. It is published inaccurately in Else Wegener's biography (1960, 75). A careful transcription of the letter, together with a commentary on its context and significance, can be found in "Alfred Wegener: Self-Proclaimed Scientific Revolutionary" (Macrakis 1984).
8. According to Hallam (1973, 68), the "germinal idea behind the theory of plate tectonics" is "clearly present" in a paper of 1965 by the Canadian geologist J. Tuzo Wilson on "transform faults," as is "the first use in this context of the term 'plates.'" But, according to Marvin (1973, 165), the "first paper to use the term 'plate' in this connection was published in 1967 by D. P. McKenzie and R. L. Parker, then at the University of California at San Diego."

9. It seems to be generally agreed that the concept of sea-floor spreading was invented by Harry Hess of Princeton in 1960 (Marvin 1973, 154–156; Hallam 1973, 54–67; Hess 1962; Cox 1973, 14–16), but not formally published by Hess until 1962, though privately circulated in 1960 in a "preprint." The first printed publication concerning sea-floor spreading was in 1961 (Dietz 1961). It has been suggested that Arthur Holmes was the "originator of [the] spreading ocean floor hypothesis" (Meyerhoff 1968). But Robert S. Dietz, in reply to the suggestion that Holmes was the "originator," said that Hess "deserves full credit for the concept. . .by reason of priority and for fully and elegantly laying down the basic premises. I have done little more than introduce the term *sea-floor spreading*. . .and apply it to such things as geosynclinal theory and the tectonic passiveness of the continental plates" (Dietz 1968; see Dietz 1963; 1966). Hess replied to the suggestion by pointing out that the "idea of sea-floor spreading was derived largely from new geophysical and topographic data on mid-ocean ridges, none of which were available to Holmes" (Hess 1968). He concluded, "The cogent term 'sea-floor spreading' which so nicely summed up my concept, was coined by Dietz (1961) after he and I had discussed the proposition at length in 1960." Dietz also substituted the eclogite-basalt phase change for the ocean floors for Hess's penidotite-serpentine composition, as I am informed by Ursula Marvin; Dietz was right, basaltic right than serpentine.
10. The major lines of research that, among others, have not been mentioned include the pinpointing of the foci of earthquakes to a small number of belts girding the earth. These belts are now recognized as the fracture lines of the earth's surface, marking the boundaries of plates coming together. Another is the study of the phenomena along the sides of plates, where one plate may move along another, which are called 'transform faults'; this topic was explored by J. Tuzo Wilson and was of great significance (see Hallam 1973, 56–59; Uyeda 1978, 65–67, 74–79; Glen 1982, 304–307, 372–375).
11. The recent revolution in earth science is unusual in that many first-rate histories have been written by professional geologists

and geophysicists soon after the event. A remarkable history of the actual post-World War II revolution is W. Glen's *The Road to Jaramillo* (1982), based on extensive reading in primary sources (both manuscript and printed) plus taped interviews with most of the major figures in the revolution. Glen comes to the subject with a background in geology and experience as author of textbook, *Continental Drift and Plate Tectonics* (1975). In 1973 Allan Cox produced an anthology of papers on *Plate Tectonics and Geomagnetic Reversals*, illuminated by the historical insights of a pioneer in the field; that same year saw the publication of Ursula B. Marvin's comprehensive *Continental Drift: The Evolution of a Concept*, especially rich for the Wegenerian and pre-Wegenerian periods, and Arthur Hallam's briefer but incisive *Revolution in Earth Sciences*. Walter Sullivan's *Continents in Motion* (1974) is a readable presentation, based on wide reading in primary sources. Seiya Uyeda's *New View of the Earth* (1978) provides a clear historically oriented overview. Another valuable historical presentation is D. P. McKenzie's extensive article on "Plate Tectonics and Its Relationship to the Evolution of Ideas in the Geological Sciences" (1977). Extremely useful are the two editions of readings from *Scientific American* edited by J. Tuzo Wilson (1976; 1973), who has also written many articles with a historical component. I have greatly profited from many discussions of this subject with the later Sir Edward (Teddy) Bullard, who wrote at least three major historical articles relating to the revolution and his own involvement in it (1965, 1975, 1975a). Finally, it should be recorded that both Wegener and du Toit included historical discussions in their respective treatises, and that the records of various symposia (e.g., the ones held in Tasmania in 1958, at the Royal Society in 1964, and at the American Philosophical Society in 1976), and the anthology (1962) edited by S. K. Runcorn on *Continental Drift*, are of enormous value for anyone studying the history of this subject.

In addition to the above-mentioned works by earth scientists, at least three historians of science (or historically minded philosophers of science) have written important studies on this revolution: David Kitts, Henry Frankel, and Rachel Laudan.

Alfred Wegener's Primary Publications*

1905

Die Alphonsinischen Tafeln. 63 pages. Dissertation, Berlin.

1906

Über die Flugbahn des am 4. Jan. 1906 in Lindenberg aufgestiegenen Registrierballons. *Beiträge zur Physik der freien Atmosphäre* 2, 30–34.

Studien über die Luftwogen. *Beiträge zur Physik der freien Atmosphäre* 2, 55–72.

1909

Die Ergebnisse der Danmark Expedition. *Gerlands Beiträge zur Geophysik* 10, 22–27.

Drachen und Fesselballonaufstiege [Danmark Expedition 1906–1908]. *Meddelelser om Grønland* 42, 1–75.

---

*A complete listing of Wegener's 170 publications can be found in Hans Benndorf, *Gerlands Beiträge zur Geophysik* 31, 337–377, 1931.

1910

Zur Schichtung der Atmosphäre. *Beiträge zur Physik der freien Atmosphäre* 3, 30–39.

Über eine eigentümliche Gesetzmäßigkeit der oberen Inversion. *Beiträge zur Physik der freien Atmosphäre* 3, 206–214.

Über eine neue fundamentale Schichtgrenze der Erd-Atmosphäre. *Beiträge zur Physik der freien Atmosphäre* 3, 225–232.

Die Größe der Wolken-Elemente. *Meteorologische Zeitschrift* 27, 354–361.

Über die Eisphase des Wasserdampfes in der Atmosphäre. *Meteorologische Zeitschrift* 27, 451–459.

1911

Untersuchungen über die Natur der obersten Atmosphärenschichten. *Physik. Zeitschrift* 12, 170–178, 214–222.

(With K. Stuchtey) Die Albedo der Wolken und der Erde. *Nachr. Ges. Wiss. Göttingen, Math.-Phys. Kl. 209–235.*

Über den Ursprung der Tromben. *Meteorologische Zeitschrift* 28, 201–209.

Neue Forschungen auf dem Gebiet der atmosphärischen Physik. *Fortschrift der naturwissenschaften Forschung* (editor E. Abderhalden) 3, 1–70.

Meteorologische Beobachtungen während der Seereise 1906 und 1909 [Danmark Expedition]. *Meddelelser om Grønland* 42, 113–124.

Meteorologische Terminbeobachtungen am Danmarks-Havn [Danmark Expedition]. *Meddelelser om Grønland* 42, 125–355.

(With J. P. Koch) Die Glaciologischen Beobachtungen der Danmark-Expedition. *Meddelelser om Grønland. 46, 5–77.*

Thermodynamik der Atmosphäre. 331 pages. Leipzig (J. A. Barth).

1912

Über die Ursache der Zerrbilder bei Sonnenuntergängen. *Beiträge zur Physik der freien Atmosphäre* 4, 26–34.

Über Temperaturinversionen. *Beiträge zur Physik der freien Atmosphäre* 4, 55–65.

Die Erforschung der obersten Atmosphärenschichten. *Gerlands Beiträge zur Geophysik* 11, 104–124.

Über turbulente Bewegungen in der Atmosphäre. *Meteorologische Zeitschrift* 29, 49–59.

(With W. Brand) Meteorologische Beobachtungen der Station Pustervig [Danmark Expedition]. *Meddelelser om Grønland* 42, 451–562.

Die Entstehung der Kontinente. *Petermanns Geographische Mitteilungen* 58, 185–195, 253–256, 305–309.

Die Entstehung der Kontinente. *Geologische Rundschau* 3, 276–292.

Barometer. Luftdruck. *Handwörterbuch der Naturwissenschaften* I, 828–839; 6, 465–471.

[Translated by Wegener] J. P. Koch, Die Reise durch Island 1912. *Petermanns Geographische Mitteilungen* 58, 185–189.

1914

Beobachtungen über atmosphärische Polarisation auf der dänischen Grönland)Expedition unter Hauptmann Koch. *Sitzungsber. Gesellschaft Marburg*, 7–18.

Staubwirbel auf Island. *Meteorologische Zeitschrift* 31, 199–200.

1915

Neuere Forschungen auf dem Gebiet der Meteorologie und Geophysik. *Annalen der Hydrographie* 43, 159–168.

Zur Frage der atmosphärischen Mondgezeiten. *Meteorologische Zeitschrift* 32, 253–258.

Über den Farbenwechsel der Meteore. *Das Wetter*. (Sonderheft Assmann-Festschrift), 5 pages.

Die Entstehung der Kontinente und Ozeane. 94 pages. Braunschweig (Vieweg).

1917

Die Neben-Sonnen unter dem Horizont. *Meteorologische Zeitschrift* 34, 295–298.

Das detonierende Meteor vom 3. April 1916, 3:30 Uhr nachmittags in Kurhessen. *Schriften. Gesellschaft Marburg* 14, 1–83.

Wind- und Wasserhosen in Europa. 301 pages. Braunschweig (Vieweg).

1918

Über die planmäßige Auffindung des Meteoriten von Treysa. *Astron. Nachr.* 207, 185–190.

Elementare Theorie der atmosphärischen Spiegelungen. *Annalen der Physik* (4) 57, 203–230.

Der Farbenwechsel großer Meteore. *Nova Acta Leop.* 104, 1–34.

1919

[Translated by Wegener] J. P. Koch, *Durch die weiße Wüste*. 247 pages, Berlin (J. Springer).

1920

Versuche zur Aufsturz-Theorie der Mondkrater. *Nova Acta Leop.* 106, 109–117.

Die Entstehung der Kontinente und Ozeane (second edition). 135 pages. Braunschweig (Vieweg).

1921

Die Theorie der Kontinentalverschiebungen. *Zeitschrift Gesellschaft Erdk. Berlin* 89–103, 125–130.

Die Entstehung der Mondkrater. 48 pages. Braunschweig (Vieweg).

1922

(With E. Kuhlbrodt) Pilotballonaufstiege auf einer Fahrt nach Mexiko, März bis Juni 1922. *Archiv der Deutschen Seewarte* 30, 1–46.

Die Entstehung der Kontinente und Ozeane (third edition). 144 pages. Braunschweig (Vieweg).

1924

Die Theorie der Kontinentverschiebung, ihr gegenwärtiger Stand und ihre Bedeutung für die exakten und systematischen Geo-Wissenschaften. *Naturwissenschaften Monatshefte* 5, 142–153.

(With W. Köppen) *Die Klimate der geologischen Vorzeit.* 256 pages. Berlin (Borntraeger).

Thermodynamik der Atmosphäre (second edition), 331 pages. Leipzig (J. A. Barth).

1925

Theorie der Haupthalos. *Archiv der Deutschen Seewarte* 43, 1–32.

Die äußere Hörbarkeitsgrenze. *Zeitschrift für Geophysik* 1, 297–314.

1926

Ergebnisse der dynamischen Meteorologie. *Ergebnisse exakt. Naturwissenschaften* 5, 96–124.

Paläogeographische Darstellung der Theorie der Kontinentalverschiebungen. *Encyklop. der Erdkd.* 174–189. Leipzig)Wien (F. Deuticke).

Thermodynamik der Atmosphäre. *Handbook der Physik* (editor H. Geiger and K. Scheel), 156–189. Berlin (J. Springer).

1927

Die geophysikalischen Grundlagen der Theorie der Kontinentverschiebung. *Scientia* 41, 102–116.

Der Boden des Atlantischen Ozeans. *Gerlands Beiträge zur Geophysik* 17, 311–321.

1928

Die Windhose in der Oststeiermark vom 23. Sept. 1927. *Meteorologische Zeitschrift* 45, 41–49.

Beiträge zur Mechanik der Tromben und Tornados. *Meteorologische Zeitschrift* 45, 201–214.

(With E. Kraus and R. Meyer) Untersuchungen über den Krater von Sall auf Ösel. *Gerlands Beiträge zur Geophysik* 20, 312–378, 428–429.

Two notes concerning my theory of continental drift. In: *The theory of continental drift* (American Association of Petroleum Geologists, Tulsa), 97–103.

Akustik der Atmosphäre, Optik der Atmosphäre. *Lehrbuch der Physik* (eleventh edition, editor Müler-Pouillet). V. I. 171–198, 199–289, Braunschweig (Vieweg).

(With J. P. Koch) Wissenschaftliche Ergebnisse der dänischen Expedition nach Dronning Louises-Land und quer über das Inlandeis von Nordgrönland 1912–13 unter Leitung von Hauptmann J. P. Koch. *Meddelelser om Grønland* 75, 676 pages.

Thermodynamik der Atmosphäre (third edition). 331 pages. Leipzig (J. A. Barth).

1929

Die Entstehung der Kontinente und Ozeane (fourth edition). 231 pages. Braunschweig (Vieweg).

Optik der Atmosphäre. B. Atmosphärische Strahlenbrechung, optische Erscheinungen in den Wolken. Lehrbuch der Geophysik (editor B. Gutenberg), 693–729. Berlin (Borntraeger).

1930

Deutsche Inlandeis-Expedition nach Grönland, Sommer 1929. *Zeitschrift der Gesellschaft für Erdkunde, Berlin, 81–124.*

Mit Motorboot und Schlitten in Grönland. 192 pages. Bielefeld and Leipzig (Velhagen and Klasing).

1935

(Edited by Kurt Wegener) *Vorlesungen über Physik der Atmosphäre*, 482 pages. Leipzig (J. A. Barth).

1961

Tagebuch eines Abenteuers. Mit Pferdeschlitten quer durch Grönland, Preface by Else Wegener. 157 pages. Weisbaden (F. A. Brockhaus).

Selected Bibliography on Wegener and Continental Drift

Benndorf, Hans. 1931. *Gerlands Beiträge zur Geophysik* 31, 337–377.
This memorial notice includes a complete listing of Wegener's 170 publications.

Büdel, J. 1978. *Die Großen der Weltgeschichte, Kindler-Zürich* 11: 460–467.
Short biography of Wegener.

Cohen, I. B. 1985. *Revolution in Science*. Cambridge: Harvard University Press, Belknap Press.

Cox, Allan. 1973. *Plate Tectonics and Geomagnetic Reversals*. San Francisco: W. H. Freeman.

Daly, Reginald. 1926. *Our Mobile Earth*. New York: Scribner.

Du Toit, Alexander. 1937. *Our Wandering Continents*. Edinburgh: Oliver and Boyd.

Frankel, Henry. 1984. "Biogeography Before and After the Rise of Sea Floor Spreading." *Studies in History and Philosophy of Science* 15, 141–168.

Frankel, Henry. 1985. "The Continental Drift Debate." In *The Resolution of Scientific Controversies: Theoretical Perspectives on Closure*, edited by A. Caplan and H. T. Englehart, Jr., 312–373. Cambridge: Cambridge University Press.

Georgi, Johannes. 1935. *Mid-Ice: The Story of the Wegener Expedition to Greenland*. New York: E. P. Dutton and Co.
Fascinating report of the hardships experienced on the 1930–1931 expedition.

Georgi, Johannes. 1960. "A. Wegener zum 80. Geburtstag." *Polarforschung* 2, Supplement, 1–102.
Reminiscences by Wegener's friend and colleague.

Glen, William. 1982. *The Road to Jaramillo*. Stanford: Stanford University Press.
Traces the development of the plate tectonic theory and the research which led to its acceptance.

Greene, Mott. Autumn, 1984. "Alfred Wegener." *Social Research* 51, no. 3: 739–761.
Clear, concise biography and history of continental drift theory.

Hallam, Anthony. 1973. *A Revolution in the Earth Sciences: from Continental Drift to Plate Tectonics*. Oxford: Clarendon Press.

Hallam, Anthony. 1975. "Alfred Wegener and the Hypothesis of Continental Drift." *Scientific American* 232, no. 2: 88–97.

Hallam, Anthony. 1983. *Great Geological Controversies*. Oxford: Oxford University Press.

Hess, H. H. 1962. "History of Ocean Basins." In *Petrologic Studies—a Volume in Honor of A. F. Buddington*, edited by A. E. J. Engel et al., 599–620. New York: Geological Society of America.

Jeffreys, Harold. 1952. *The Earth: Its Origin, History and Physical Constitution*. Cambridge: Cambridge University Press.

Köppen, Wladimir and Alfred Wegener. 1924. *Die Klimate der Geologischen Vorzeit*. Berlin: Bonntraeger.

Köppen, Wladimir and Alfred Wegener. 1963. *Climates of the Past*. London: D. van Nostrand.

Marvin, Ursula. 1973. *Continental Drift: The Evolution of a Concept*. Washington: Smithsonian Institution Press.

McKenzie, D. P. 1977. "Plate Tectonics and Its Relationship to the Evolution of Ideas in the Geological Sciences." *Daedalus* 106: 97–124.

Runcorn, S. K., editor. 1962. *Continental Drift*. New York: Academic Press.

Schmauss, August. 1951. "Alfred Wegeners Leben und Wirken als Meteorologie." *Annalen der Meteorologie* 4: 1–13.

Sullivan, Walter. 1974. *Continents in Motion: The New Earth Debate*. New York: McGraw-Hill.
A well-written, popular account.

Totten, Stanley M. 1981. "Frank B. Taylor, Plate Tectonics, and Continental Drift." *Journal of Geological Education* 29: 212–220.

Van Waterschoot Van der Gracht, W. A. J. M., editor. 1928. *Theory of Continental Drift: A Symposium*. Tulsa: American Association of Petroleum Geologists.
Report of the 1926 meeting in New York. Includes a paper by Wegener, pp. 97–103.

Wegener, Alfred. 1924. *The Origin of Continents and Oceans*. Translated from the third edition. London: Methuen.

Wegener, Alfred. 1966. *The Origin of Continents and Oceans*. Translated from the fourth edition. New York: Dover Publications.

Wegener, Else, editor. 1939. *Greenland Journal*. London and Glasgow: Blackie and Son.

Wegener, Else, editor. 1932. *Alfred Wegeners letzte Grönlandfahrt*. Leipzig. F. A. Brockhaus.

Wegener, Else. 1960. *Alfred Wegener. Tagebücher, Briefe, Erinnerungen*. Wiesbaden: F. A. Brockhaus.

Wilson, J. Tuzo. 1976. *Continents Adrift and Continents Aground*. San Francisco: W. H. Freeman and Co.

Figure Acknowledgments

Page viii Dr. Olbers—Hamburg. *Page xv* Heimatsmuseum Neuruppin, from the Else Wegener Archive. *Page 11* M. Schwarzbach. *Page 12* M. Schwarzbach. *Page 18* Heimatsmuseum Neuruppin. *Page 20* Deutsches Museum, Munich. *Page 21* M. Schwarzbach. *Page 23* Deutsches Museum, Munich. *Page 26* M. Schwarzbach. *Page 34* Deutsches Museum, Munich. *Page 36* From *Alfred Wegeners Letzte Grönlandfahrt* by Else Wegener (AWLG). *Page 37* AWLG. *Page 38* Alfred Wegener Institut für Polarforschung (IfP). *Page 39* IfP. *Page 41* Deutsches Museum, Munich. *Page 45* IfP. *Page 46* IfP. *Page 47* AWLG. *Page 48* IfP. *Page 49* *Mid-Ice* by J. Georgi (MI). *Page 51* Deutsches Museum, Munich. *Page 52* MI. *Page 54* IfP. *Page 56* *Die weisse Wüste* by K. Herdemerten. *Page 62* *Alfred Wegener* by Else Wegener. *Page 63* *Die weisse Wüste* by K. Herdemerten. *Page 65* M. Schwarzbach. *Page 67* *Die Entstehung der Kontinente und Ozeane*, 1920, by Alfred Wegener. *Page 78* A.V. Carozzi, *Compt. rend. soc. phys.* Geneva, 4, 171, 1969. *Page 82* A. Wegener, *Geologisches Rundschau*, 3, 276, 1912. *Page 87* E. Kuhlbrodt. *Page 91* *Die Klimate der geologischen Vorzeit*, by W. Köppen and A. Wegener. *Page 93* *Die Klimate der geologischen Vorzeit*, by W. Köppen and A. Wegener. *Page 96* K. Brunnacker. *Page 97* *Die Klimate der geologischen Vorzeit*, by W. Köppen and A. Wegener. *Page 114* Redrawn from O. Ampferer, *Die Naturwissenschaften*, 31, 1925. *Page 116* S.H. Haughton, *Proc. Geol. Soc. Am.* 141, 1949. *Page 129* Redrawn from Dietz and Holden, *Jour. Geophys. Res.* 75, 1970. *Page 137* M. Koglbauer. *Page 138* M. Schwarzbach. *Page 157* MI. *Page 165* MI. *Page 183* Deutsches Museum, Munich.

INDEX